Concise
Encyclopedia of
Robotics

Other great robotics titles from TAB Electronics:

Concise
Encyclopedia of
Robotics

Stan Gibilisco

McGraw-Hill
New York   Chicago   San Francisco   Lisbon   London   Madrid
Mexico City   Milan   New Delhi   San Juan   Seoul
Singapore   Sydney   Toronto

The *McGraw·Hill* Companies

Library of Congress Cataloging-in-Publication Data
Concise encylopedia of robotics/[edited by] Stan Gibilisco
    p.   cm.
    Includes index.
    ISBN 0-07-141010-4
    1. Robotics—Encyclopedias.   2. Artificial intelligence—Encyclopedias.   3. Robotics
I. Gibilisco, Stan.

TJ210.4.C65   2002
629.8'92—dc21   2002032000

1 2 3 4 5 6 7 8 9 0   DOC/DOC   0 9 8 7 6 5 4 3 2

ISBN 0-07-141010-4

The sponsoring editor for this book was Scott Grillo, the editing supervisor was David E. Fogarty, and the production supervisor was Sherri Souffrance. It was set in Minion by Wayne Palmer and Kim Sheran of McGraw-Hill's Professional's composition unit, Hightstown, N.J.

Printed and bound by RR Donnelley.

McGraw-Hill books are available at special quantity discounts to use as premiums and sales promotions, or for use in corporate training programs. For more information, please write to the Director of Special Sales, Professional Publishing, McGraw-Hill, Two Penn Plaza, New York, NY 10121-2298. Or contact your local bookstore.

To Samuel, Tim, and Tony
from Uncle Stan

Contents

Foreword   ix

Introduction   xi

Acknowledgments   xiii

Concise Encyclopedia of Robotics and AI   1

Suggested Additional References   351

Index   353

Foreword

Welcome to the high frontier of cognition and computation!

You are about to dig into a uniquely interesting and important book. This is not a highly technical or abstruse guide to this often complex and difficult-to-understand subject. Rather, the book provides short, clear definitions and interpretations of the major concepts and rapidly emerging ideas in this dynamic field. The work includes numerous functional illustrations that help the general reader "see" the abstract robotics notions presented. I envision this book as an important introductory overview for the general interested reader and artificial intelligence (AI) hobbyist, and as a valuable backup and refresher for professional workers in the area. Because this book is all about terms, let me frame the effort by saying that it addresses two major, universal needs: *cognition* and *computation*.

Cognition (kog-NISH-un) is literally "the act of knowing or awareness." As you refer to definitions in this book, you should steadily gain knowledge and awareness of the robotics/AI topics presented. Most interestingly, it is your own *central nervous system* (brain and spinal cord), functionally connected to your eyes, that is allowing your *natural intelligence (NI)* to study this material and build a relevant cognition (mental awareness) of key robotics concepts. A thorough cognition and comprehension of robotics/AI terminology and concepts is becoming absolutely critical to all intelligent lay people worldwide.

"Can human consciousness be duplicated electronically?" Stan Gibilisco thoughtfully asks us. "Will robots and smart machines ever present a danger to their makers? What can we reasonably expect from robotics and artificial intelligence in the next 10 years? In 50 years? In 100 years?" Such questions as these will be of ever-increasing importance for human cognition (NI) as machine computation or artificial intelligence continues to evolve rapidly.

Let us coin a new word, *computhink*: a contraction of "*computer*-like modes or ways of human *thinking*." A comprehensive introductory reference book such as this will help the NI of general readers to learn computhink. This will result in a better understanding and management of our powerful cousin, AI, for the greater benefit and education of all humankind.

This book presents a thorough, basic, blissfully nonmathematical coverage of numerous electronic and mechanical concepts that is greatly needed worldwide. Stan has provided us with the essential vocabulary of machine computation for the twenty-first century, a vocabulary that many (not just a select few) people need to understand.

THE HONORABLE DR. DALE PIERRE LAYMAN, PH.D.
Founder and President, ROBOWATCH
www.robowatch.org

Introduction

This is an alphabetical reference about robotics and artificial intelligence (AI) for hobbyists, students, and people who are just curious about these technologies.

Computers and robots are here to stay. We depend on them every day. Often we don't notice them until they break down. We will get more used to them, and more reliant on them, as the future unfolds.

To find information on a subject, look for it as an article title. If your subject is not an article title, look for it in the index.

This book is meant to be precise, but without too much math or jargon. It is written with one eye on today and the other eye on tomorrow. Illustrations are functional; they are drawn with the intention of showing, clearly and simply, how things work.

Suggestions for future editions are welcome.

STAN GIBILISCO
Editor in Chief

Acknowledgments

Illustrations in this book were generated with CorelDRAW. Some clip art is courtesy of Corel Corporation, 1600 Carling Avenue, Ottawa, Ontario, Canada K1Z 8R7.

A

ACOUSTIC PROXIMITY SENSOR

An *acoustic proximity sensor* can be used by a robot to detect the presence of, and determine the distance to, an object or barrier at close range. It works based on acoustic wave interference. The principle is similar to that of *sonar*, but rather than measuring the time delay between the transmission of a pulse and its echo, the system analyzes the phase relationship between the transmitted wave and the reflected wave.

When an acoustic signal having a single, well-defined, constant frequency (and therefore a single, well-defined, constant wavelength) reflects from a nearby object, the reflected wave combines with the incident wave to form alternating zones at which the acoustic energy adds and cancels in phase. If the robot and the object are both stationary, these zones remain fixed. Because of this, the zones are called *standing waves*. If the robot moves with respect to the object, the standing waves change position. Even a tiny shift in the relative position of the robot and the sensed object can produce a considerable change in the pattern of standing waves. This effect becomes more pronounced as the acoustic wave frequency increases, because the wavelength is inversely proportional to the frequency.

The characteristics and effectiveness of an acoustic proximity sensor depend on how well the object or barrier reflects acoustic waves. A solid concrete wall is more easily detected than a sofa upholstered with cloth. The distance between the robot and the obstacle is a factor; in general, an acoustic proximity sensor works better as the distance decreases, and less well as the distance increases. The amount of *acoustic noise* in the robot's work environment is also important. The higher the noise level, the more limited is the range over which the sensor functions, and the more likely are errors or false positives. Ultrasound waves provide exceptional accuracy at close range, in some cases less than 1 cm. Audible sound can allow the system to function at distances on the order of several meters. However, audible signals can annoy people who must work around the machine. Compare SONAR.

See also PRESENCE SENSING and PROXIMITY SENSING.

ACTIVE CHORD MECHANISM (ACM)

An *active chord mechanism (ACM)* is a robot gripper that conforms to the shapes of irregular objects. An ACM is built something like the human backbone. A typical ACM consists of numerous small, rigid structures connected by hinges, as shown in the illustration.

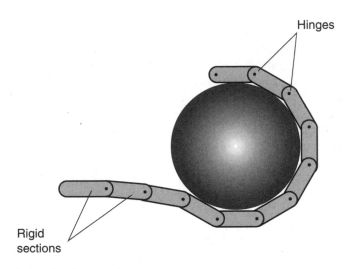

Hinges

Rigid sections

Active chord mechanism

The precision with which an ACM can conform to an irregular object depends on the size and number of sections. The smaller the sections, the greater is the precision. An ACM exerts uniform pressure all along its length. This pressure can be increased or decreased, according to the required task.

One application of ACMs is to position or arrange fragile objects without damaging them. Another application is the picking of fruits and vegetables.

See also ROBOT GRIPPER.

ACTIVE COOPERATION

See COOPERATION.

ACTUATOR

An *actuator* is a device that moves one or more joints and operates the gripper or end effector in a *robot arm*. Simple actuators consist of electric motors and gears, *cable drives,* or *chain drives.* More sophisticated actuators involve the use of hydraulics, pneumatics, or magnetic interaction. *Stepper motors* are commonly used as robotic actuators.

Some robot arms can function with a single actuator; others require two or more. The number of actuators necessary to perform a given task depends on the number of degrees of freedom, the number of degrees of rotation, and the coordinate geometry of the robot arm.

See also CABLE DRIVE, CHAIN DRIVE, DEGREES OF FREEDOM, DEGREES OF ROTATION, END EFFECTOR, MOTOR, ROBOT ARM, ROBOT GRIPPER, and STEPPER MOTOR.

ADAPTIVE SUSPENSION VEHICLE (ASV)

An *adaptive suspension vehicle (ASV)* is a specialized robot that uses mechanical limbs to propel itself. It moves on several legs like a gigantic insect. This provides excellent stability and maneuverability. The ASV can carry several hundred kilograms, and moves at 2 to 4 m/s. The machine itself masses 2 to 3 metric tons. It is the size of a small truck, and it can carry a driver or rider.

The design and construction of a robot with legs is considerably more difficult than that of a wheel-driven or track-driven robot, but there is a payoff: the ASV can move over much rougher terrain than any vehicle with wheels or a track drive.

See also INSECT ROBOT and ROBOT LEG.

ADHESION GRIPPER

An *adhesion gripper* is a robot *end effector* that grasps objects by literally sticking to them. In its most primitive form, this type of gripper consists of a rod, sphere, or other solid object covered with two-sided tape. Velcro™ can also be used if the object(s) to be grasped are likewise equipped.

A major asset of the adhesive gripper is the fact that it is simple. As long as the adhesive keeps its "stickiness," it will continue to function without maintenance. However, there are certain limitations. The most significant is the fact that the adhesive cannot readily be disabled in order to release the grasp on an object. Some other means, such as devices that lock the gripped object into place, must be used. Compare ATTRACTION GRIPPER.

AGV

See AUTOMATED GUIDED VEHICLE.

ALGORITHM

An *algorithm* is a precise, step-by-step procedure by which a solution to a problem is found. Algorithms can usually be shown in flowchart form. All computer programs are algorithms. Robots perform specific tasks by following algorithms that tell them exactly where and when to move.

In an efficient algorithm, every step is vital, even if it seems to sidetrack or backtrack. An algorithm must contain a finite number of steps. Each step must be expressible in digital terms, allowing a computer to execute it. Although the algorithm can contain loops that are iterated many times, the whole process must be executable in a finite length of time. Although no algorithm is infinitely complex, there are some that would require millions of years to be executed by a human being but can be done by computers in a few seconds.

See also FLOWCHART.

ALL-TRANSLATIONAL SYSTEM

An *all-translational system* is a scheme in which the coordinate axes remain constant, or fixed, in an absolute sense as a robot moves. A common example is a system in three-dimensional (3-D) *Cartesian coordinate geometry*, in which the axes are defined as north/south, east/west, and up/down.

An all-translational system in a given environment does not necessarily constitute an all-translational system in another environment. Consider a Cartesian system in which the x axis is north/south, the y axis is east/west, and the z axis is up/down. This is all-translational as defined in, and relative to, a small region on the Earth. However, this scheme loses its absoluteness with respect to the whole planet or the greater Universe, because the Earth is a rotating sphere, not a fixed Euclidean plane.

In the absence of a set of physical objects for reference, an all-translational system can be maintained by inertial means. The *gyroscope* is the most common means of accomplishing this.

See also CARTESIAN COORDINATE GEOMETRY and GYROSCOPE.

ALTERNATIVE COMPUTER TECHNOLOGY

Researchers in artificial intelligence (AI) have debated for years whether it is possible to build a machine with intelligence comparable to that of a human being. Some scientists think that *alternative computer technology* might provide a pathway in the quest for human-level AI.

Digital processes

Personal computers make use of digital computer technology. The operating language, known as *machine language*, consists of only two possible states, the digits 1 and 0, represented by high and low electronic voltages. No matter how complex the function, graphic, or program, the workings of a digital computer can always be broken down into these two logic states.

Digital computers can be made fast and powerful. They can work with huge amounts of data, processing it at many millions of digits per second. However, there are certain things that digital computers are not good at doing. Some researchers think that other approaches to computing deserve attention, even though digital technology has been successful so far.

Analog processes

Whereas a digital machine breaks everything down into discrete bits (binary digits), *analog computer technology* uses an entirely different approach. Think of the square root of 2. This cannot be represented as a ratio of whole numbers. A digital computer will calculate this and get a value of about 1.414. However, a decimal-number representation of the square root of 2 can never be exact. The best a digital machine can do is get close to its true value.

The square root of 2 is the length of the diagonal of a square measuring 1 unit on a side. You can construct it with the tools of classical geometry (an analog art) and get an exact rendition. But you cannot use this in arithmetic as you use the numerical value 1.414. Thus, you sacrifice quantitative utility for qualitative perfection. Perhaps similar give-and-take will prove necessary in the quest to develop a computer that thinks like a human being. Analog concepts have been adapted to computer design; in fact, it was one of the earliest methods of computing. In recent years it has been largely ignored.

Optics

Visible light, infrared (IR), and ultraviolet (UV) offer interesting possibilities for the future of computer technology.

In CD-ROM (compact disk, read-only memory), optical technology is used to increase the amount of data that can be stored in a given physical space. Tiny pits on a plastic diskette cause a laser beam to be reflected or absorbed at the surface. This allows encoding of many megabytes of data on a diskette less than 15 cm across.

Data can be transmitted at extreme speeds, and in multiple channels, via lasers in glass fibers. This is known as *fiber-optic data transmission,* and is used in some telephone systems today. The wires in computers might someday be replaced by optical fibers. The digital logic states, now represented by electrical impulses or magnetic fields, would be represented by light transmittivity instead. Certain materials change their optical properties very quickly, and can hold a given state for a long time.

Atomic data

As *integrated circuit (IC)* technology has advanced, more and more digital logic gates have been packed into less and less physical space. Also, with refinements in magnetic media, the capacity of hard disks and diskettes has been increasing.

According to conventional science, the smallest possible data storage unit is a single atom or subatomic particle. Consider a magnetic diskette. Logic 1 might be represented by an atom "right side up," with its magnetic north pole facing upward and its magnetic south pole facing downward. Then logic 0 would be represented by the same atom "upside down," with the magnetic poles inverted.

Another possibility is *single-electron memory (SEM)*. An example of a SEM is a substance in which the presence of an excess electron in an atom represents logic 1, and the electrically neutral state of the atom represents logic 0.

Some scientists think that computer chips might someday be grown in a laboratory, in a manner similar to the way experimental cultures of bacteria and viruses are grown. A name has even been coined for such a device: *biochip.*

Nanotechnology

As ICs get more circuitry packed into small packages, computer power increases. But it also becomes possible to make tinier and tinier computers. With *molecular computer technology*—the construction of ICs molecule by molecule rather than by etching material away from a chip—it might become possible to build computers so small that they can circulate inside the human body.

Imagine antibody robots, controlled by a central computer, that are as small as bacteria. Suppose the central computer is programmed to destroy certain disease-causing organisms. Such a machine would be something like an artificial white blood cell. Nanotechnology is the field of research devoted to the development and programming of microscopic machines. The prefix *nano-* means one billionth (10^{-9} or 0.000000001). It also means "extremely small."

Computerized *nanorobots* might assemble larger computers, saving humans much of the work now associated with manufacturing the machines. Nanotechnology has already made it possible for you to wear a computer on your wrist, or even have a computer embedded somewhere in your body.

See also BIOCHIP, INTEGRATED CIRCUIT, NANOCHIP, and NEURAL NETWORK.

Neural networks

Neural network technology uses a design philosophy that differs radically from that of conventional digital computers. Neural networks are good at spotting patterns, which is important for forecasting. Rather than working with discrete binary digits, neural networks work with the relationships among events.

Unless there is a malfunction, a digital machine does precise things with data. This takes time, but the outcome is always the same if the input stays constant. This is not the case with a neural network. A neural network can work more quickly than a digital machine. To achieve speed, precision is sacrificed. Neural networks can learn from their mistakes. According to some scientists, this technology is a diversion and distraction from the proven mainstream; according to other scientists, it holds great promise.

AMUSEMENT ROBOT

An *amusement robot* is a hobby robot intended for entertainment or gaming. Companies sometimes use them to show off new products and to attract customers. They are common at trade fairs, especially in Japan. Although they are usually small in size, they often have sophisticated controllers.

An example of an amusement robot is a mechanical mouse (not to be confused with the pointing device for a computer) that navigates a maze. The simplest such device bumps around randomly until it finds its way out by accident. A more sophisticated robot mouse moves along one wall of the maze until it emerges. This technique will work with most, but not all, mazes.

The most advanced amusement robots include *androids,* or machines with a human appearance. Robots of this type can greet customers in stores, operate elevators, or demonstrate products at conventions. Some amusement robots can accommodate human riders.

See also ANDROID and PERSONAL ROBOT.

ANALOGICAL MOTION

The term *analogical motion* refers to a variable or quantity that can have an infinite number of values within a certain range. This is in contrast to digital variables or quantities, which can have only a finite number of discrete values within a given range. Thus, analogical control is representative of so-called smooth or continuous motion.

A person moving freely around a room, varying position to any point within a specific region, has the capability of analogical motion. The human arm can move to an infinite number of positions in a fluid and continuous way, within a certain region of space. This, too, is analog motion. Many robots, however, can move only to certain points along a line, on a plane, or in space. This motion is digital. Some robots can move in an analogical fashion, but the necessary hardware is generally more complicated than for digital motion.

The illustration shows an example of analog motion in a plane. Compare DIGITAL MOTION.

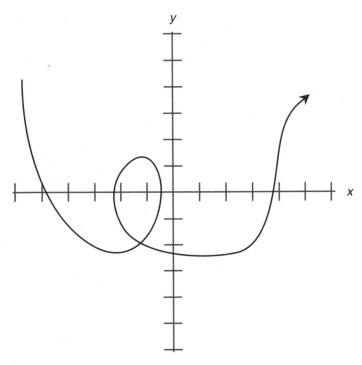

Analogical motion

ANALYTICAL ENGINE

The *analytical engine* was a primitive calculating machine designed by Charles Babbage in the nineteenth century. Babbage never completed the task of building this device to perfection, but the idea was to employ punched cards to make, and print out, calculations, in a manner similar to the earliest digital computers. Babbage is considered to be the first engineer to work on a true digital calculator.

One of the main problems for Babbage was that electricity was not available. The machines had to use mechanical parts exclusively. These wore out with frequent, repetitive use. Another problem was that Babbage liked to dismantle things completely in order to start over with new designs, rather than saving his old machines to keep their shortcomings in mind when designing new ones.

During the research-and-development phase of the analytical engine, some people thought that *artificial intelligence (AI)* had been discovered. The Countess of Lovelace even went so far as to write a program for the machine. Babbage's machine represented a turning point in human attitudes toward machines. People began to believe that "smart machines" were not only possible in theory, but also practical.

AND GATE

See LOGIC GATE.

ANDROID

An *android* is a robot that has human form. A typical android has a rotatable head equipped with position sensors. *Binocular machine vision* allows the android to perceive depth, thereby locating objects anywhere within a large room. *Speech recognition* and *speech synthesis* can be included as well. Because of their quasi-human appearance, androids are especially suited for use where there are children.

There are certain mechanical problems with design of humanoid robots. *Biped robots* are unstable. Even three-legged designs, while more stable, are two-legged whenever one of the legs is off the ground. Humans have an innate sense of balance, but this feature is difficult to program into a machine. Thus, an android usually propels itself by means of a *wheel drive* or *track drive* in its base. Elevators can be used to allow a rolling android to get from floor to floor in a building.

The technology exists for fully functional arms, but the programming needed for their operation has not yet been made cost-effective for small robots.

No android has yet been conceived, even on the trendiest drawing board, that can be mistaken for a person, as has been depicted in science-fiction books and movies.

Humanoid robots have enjoyed popularity, especially in Japan. One of the most famous was called *Wasubot*. It played an organ with the finesse of a professional musician. This robot became an idol at the Japanese show Expo '85. The demonstration showed that machines can be esthetically appealing as well as functional.

See also PERSONAL ROBOT.

ANIMISM

People in some countries, notably Japan, believe that the force of life exists in things such as stones, lakes, and clouds, as well as in people, animals, and plants. This belief is called *animism*.

As early as the middle of the nineteenth century, a machine was conceived that was thought to be in some sense animate. This was Charles Babbage's *analytical engine*. At that time, very few people seriously thought that a contraption made of wheels and gears could have life. However, today's massive computers, and the promise of more sophisticated ones being built every year, have brought the question out of the realm of science fiction.

Computers can do things that people cannot. For example, even a simple personal computer (PC) can figure out the value of π (pi), the ratio of a circle's circumference to its diameter, to millions of decimal places. Robots can be programmed to do things as complicated as figuring out how to get through a maze or rescue a person from a burning building. In recent years, programming has progressed to the point that computers can learn from their mistakes, so that they do not make any particular error more than once. This is one of the criteria for intelligence, but few Western engineers or scientists consider this, by itself, characteristic of life.

ANTHROPOMORPHISM

Sometimes, machines or other objects have characteristics that seem human-like to us. This is especially true of advanced computers and robots. We commit *anthropomorphism* when we think of a computer or robot as human. Androids, for example, are easy to anthropomorphize. Science-fiction movies and novels often make use of anthropomorphisms.

An example of anthropomorphism with respect to a computer occurs in the novel and movie *2001: A Space Odyssey*. In this story, a spacecraft is controlled by "Hal," a computer that becomes delusional and tries to kill the human astronauts.

Some engineers believe that sophisticated robots and computers already have human qualities, because they can optimize problems and/or learn from their mistakes. Others, however, contend that the criteria for life are far more strict.

Owners of personal robots sometimes think of the machines as companions. In that sense, such robots actually are like people, because it is possible to grow fond of them.

See also PERSONAL ROBOT.

ARM

See ROBOT ARM.

ARTICULATED GEOMETRY

Robot arms can move in various different ways. Some can attain only certain discrete, or definite, positions, and cannot stop at any intermediate position. Others can move in smooth, sweeping motions, and are capable of reaching to any point within a certain region.

One method of robot arm movement is called *articulated geometry*. The word "articulated" means "broken into sections by joints." This type of robot arm resembles the arm of a human. The versatility is defined in terms of the number of *degrees of freedom*. There might, for example, be base rotation, elevation, and reach. There are several different articulated geometries for any given number of degrees of freedom. The illustration shows one scheme for a robot arm that uses articulated geometry.

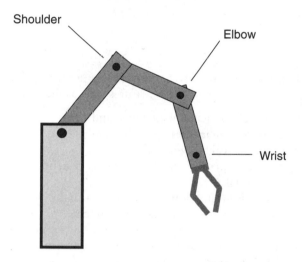

Articulated geometry

Other geometries that facilitate movement in two or three dimensions are defined under the titles CARTESIAN COORDINATE GEOMETRY, CYLINDRICAL COORDINATE GEOMETRY, DEGREES OF FREEDOM, POLAR COORDINATE GEOMETRY, and SPHERICAL COORDINATE GEOMETRY.

ARTIFICIAL INTELLIGENCE

The definition of what constitutes *artificial intelligence (AI)* varies among engineers. There is no universally accepted agreement on its exact meaning.

The programming of robots can be divided into levels, starting with the least sophisticated and progressing to the theoretical, rather nebulous level of AI. The drawing shows a four-level programming scheme.

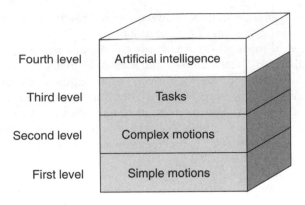

Fourth level	Artificial intelligence
Third level	Tasks
Second level	Complex motions
First level	Simple motions

Artificial intelligence

Artificial intelligence, at the top level, encompasses properties, behaviors, and tasks and involves robots with features such as the ability to:

- Sense physical variables such as light and sound
- Generate high-resolution images (vision system)
- Develop a concept of reality (world model)
- Determine the optimum or most efficient course of action
- Learn from past mistakes
- Create a plan in a given situation, and then follow it through
- Modify a plan as changes occur in the environment
- Carry on two-way conversations with humans or other machines
- Infer solutions based on limited or incomplete information
- Develop new ways to solve old problems
- Search the knowledge base for specific facts or solutions
- Program themselves
- Improve their own designs

Artificial intelligence is difficult to quantify; the most tempting standard is to compare "machine intelligence" with human intelligence. For example, a smart machine can be given an intelligence quotient (IQ) test similar to the tests designed to measure human intelligence. In this interpretation, the level of AI increases as a robot or computer becomes more "human-like" in its reactions to the world around it. Another scheme involves the use of games requiring look-ahead strategy, such as checkers or chess.

ARTIFICIAL STIMULUS

An *artificial stimulus* is a method of guiding a robot along a specified path. The *automated guided vehicle* (AGV), for example, makes use of a magnetic field to follow certain routes in its environment.

Various landmarks can be used as artificial stimuli. It is not necessary to have wires or magnets embedded in the floor, as is the case with the AGV. A robot might be programmed to follow the wall on its right-hand (or left-hand) side until it reaches its destination, like finding its way out of a maze. The lamps in a hallway ceiling can be followed by light and direction sensors. The edge of a roadway can be followed by visually checking the difference in brightness between the road surface and the shoulder.

Another way to provide guidance is to use a beacon. This can be an infrared (IR) or visible beam, or a set of ultrasound sources. With ultrasound, the robot can measure the difference in propagation time from different sources to find its position in an open space, if there are no obstructions.

There are many ways that objects can be marked for identification. One method is *bar coding,* which is used for pricing and product identification in retail stores. Another is a *passive transponder,* of the type attached to merchandise to prevent shoplifting.

See also AUTOMATED GUIDED VEHICLE, BAR CODING, BEACON, EDGE DETECTION, and PASSIVE TRANSPONDER.

ASIMOV'S THREE LAWS

In one of his early science-fiction stories, the prolific writer *Isaac Asimov* first mentioned the word "robotics," along with three fundamental rules that all robots had to obey. The rules, now called *Asimov's three laws,* are as follows.

- A robot must not injure, or allow the injury of, any human being.
- A robot must obey all orders from humans, except orders that would contradict the First Law.
- A robot must protect itself, except when to do so would contradict the First Law or the Second Law.

Although these rules were first coined in the 1940s, they are still considered good standards for robotic behavior.

ASSEMBLY ROBOT

An *assembly robot* is any robot that assembles products, such as cars, home appliances, or electronic equipment. Some assembly robots work alone; most are used in *automated integrated manufacturing systems* (AIMS), doing repetitive work at high speed and for long periods of time.

Many assembly robots take the form of robot arms. The type of joint arrangement depends on the task that the robot must perform. Joint arrangements are named according to the type of coordinate system they follow. The complexity of motion in an assembly robot is expressed in terms of the number of *degrees of freedom.*

To do its work properly, an assembly robot must have all the parts it works with placed in exactly the correct locations. This ensures that the robot can pick up each part in the assembly process, in turn, by moving to the correct set of coordinates. There is little tolerance for error. In some assembly systems, the various components are labeled with identifying tags such as bar codes, so the robot can find each part by zeroing in on the tag.

See also CARTESIAN COORDINATE GEOMETRY, CYLINDRICAL COORDINATE GEOMETRY, DEGREES OF FREEDOM, POLAR COORDINATE GEOMETRY, ROBOT ARM, and SPHERICAL COORDINATE GEOMETRY.

ATTRACTION GRIPPER

An *attraction gripper* is a robot *end effector* that grasps objects by means of electrical or magnetic attraction. Generally, magnets are used; either permanent magnets or electromagnets will serve the purpose. Electromagnets offer the advantage of being on/off controllable, so an object can be conveniently released without its having to be secured by some external means. Permanent magnets, conversely, offer the advantage of a minimal maintenance requirement.

Like the *adhesive gripper,* the attraction gripper is fundamentally simple. There are two primary problems with this type of end effector. First, in order for a magnetic attraction gripper to work, the object it grasps must contain a ferromagnetic material such as iron or steel. Second, the magnetic field produced by the end effector can permanently magnetize the objects it handles. In some cases this is not a concern, but in other instances it can cause trouble. Compare ADHESION GRIPPER.

ATTRACTIVE RADIAL FIELD

See POTENTIAL FIELD.

AUTOMATED GUIDED VEHICLE

An *automated guided vehicle (AGV)* is a robot cart that runs without a driver. The cart has an electric engine and is guided by a magnetic field, produced by a wire on or just beneath the floor (see the illustration). Alternatively, an AGV can run on a set of rails. In automated systems, AGVs are used to bring components to assembly lines. AGVs can also serve as attendants in hospitals, bringing food and nonessential items to patients, or as mechanical gophers to perform routine chores around the home or office.

There has been some talk about making automobiles into AGVs that follow wires embedded in the road pavement. This would take part of the driver's job away, letting computers steer the vehicle and adjust its speed.

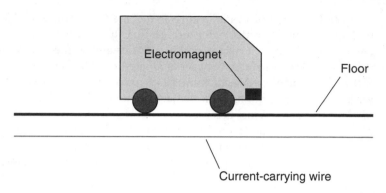

Automated guided vehicle

Each car would have its own individual computer. In a city, traffic would be overseen by one or more central computers. In the event of computer failure, all traffic would stop. This would practically eliminate accidents. Whether the public would accept this sort of system on a general scale remains to be seen.

AUTOMATED HOME

See SMART HOME.

AUTOMATION

The term *automation* refers to a system in which some or all of the processes are done by machines, particularly robots. Assets of automation include the following:

- Robots work fast.
- Robots are precise.
- Robots are reliable if they are well designed and maintained.
- Robots are capable of enormous physical strength.

Advantages of human operators over robots include these facts:

- People can solve some problems that machines cannot.
- People have greater tolerance for confusion and error.
- Humans can perform certain tasks that robots cannot.
- Humans are needed to supervise robotic systems.

AUTOMATON

An *automaton* is a simple robot that performs a task or set of tasks without sophisticated computer control. Automata have been around for over 200 years.

An early example of an automaton was the "mechanical duck" designed by J. de Vaucanson in the eighteenth century. It was used to entertain audiences in Europe. It made quacking sounds and seemed to eat and drink. Vaucanson used the robot act to raise money for his work.

Every December, certain ambitious people build holiday displays in their yards, consisting of machines in the form of people and animals. These machines have no "brains," because they simply follow mechanical routines. Although they are fun to observe, these devices lack precision, and the motions they can make are limited. Some of these machines may look like androids, but are actually no more than moving statues. Compare ANDROID.

AUTONOMOUS ROBOT

An *autonomous robot* is self-contained, housing its own *controller,* and not depending on a central computer for its commands. It navigates its work environment under its own power, usually by rolling on wheels or a track drive.

Robot autonomy might at first seem like a great asset: if a robot functions by itself in a system, then when other parts of the system fail, the robot will keep working. However, in systems where many identical robots are used, autonomy is inefficient. It is better from an economic standpoint to put programs in one central computer that controls all the robots. *Insect robots* work this way.

Simple robots, like those in assembly lines, are not autonomous. The more complex the task, and the more different things a robot must do, the more autonomy it can have. The most advanced autonomous robots have *artificial intelligence (AI).*

See also ANDROID and INSECT ROBOT.

AXIS INTERCHANGE

Axis interchange is the transposition of coordinate axis in a robotic system that uses *Cartesian coordinate geometry.* Axis interchange can involve two axes, or all three.

The illustration shows an example in which the left/right (normally x) and up/down (normally z) axes are transposed. This is not the only way in which the left/right versus up/down interchange can take place; one or both axes might also be inverted. Clearly, there are numerous possibilities for axis interchange in a three-dimensional Cartesian system.

Axis interchange can produce useful variations in robot movements. A single-motion programming scheme can result in vastly different *work envelopes* and motion patterns, depending on how the axes are defined. No matter how the axes are transposed, however, there is always a one-to-one correspondence between the points in both work envelopes, provided the motion programming is done properly.

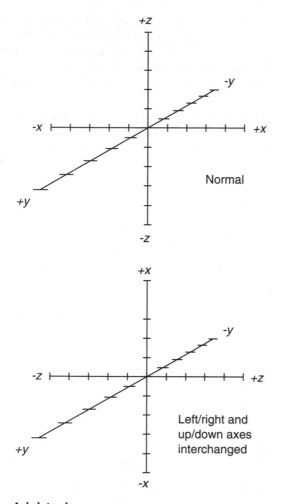

Axis interchange

Depending on the type of robotic system used, axis interchange can alter or limit the work envelope. Certain position points, or certain types of motion, that are possible in one coordinate scheme might be impossible in the other.

See also AXIS INVERSION, CARTESIAN COORDINATE GEOMETRY, and WORK ENVELOPE.

AXIS INVERSION

Axis inversion is a reversal in the orientation of one or more coordinate axes in a robotic system that uses *Cartesian coordinate geometry*.

When robotic motions are programmed using the Cartesian (or rect-angular) scheme, the difference between right-handed and left-handed operations consists only of the reversal, or inversion, of the coordinates in one of the axes. Generally, the left/right axis in a Cartesian scheme is the *x* axis. The reversal of the coordinates in this axis is a form of *single-axis inversion*.

The illustration shows two three-dimensional Cartesian coordinate grids. In the top example, a right-handed scheme is depicted. The lower drawing shows the left-handed equivalent. The coordinate designations

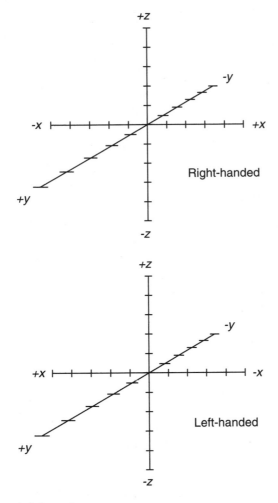

Axis inversion

are identical, except that they are mirror images with respect to the x axis. All the divisions represent the same unit distance in either case. While left and right are reversed in this example, the senses of up/down and forward/backward remain the same.

In some systems, it is necessary to invert two, or even all three, axes to obtain the desired robot motion. These schemes can be called *dual-axis inversion* or *triple-axis inversion*.

See also AXIS INTERCHANGE and CARTESIAN COORDINATE GEOMETRY.

AZIMUTH-RANGE NAVIGATION

Electromagnetic (EM) or acoustic waves reflect from various objects. By ascertaining the directions from which transmitted EM or acoustic signals are returned, and by measuring the time it takes for pulses to travel from the transmitter location to a target and back, it is possible for a robot to locate objects within its work environment. The ongoing changes in the azimuth (compass bearing) and range (distance) information for each object in the work environment can be used by the robot controller for navigation.

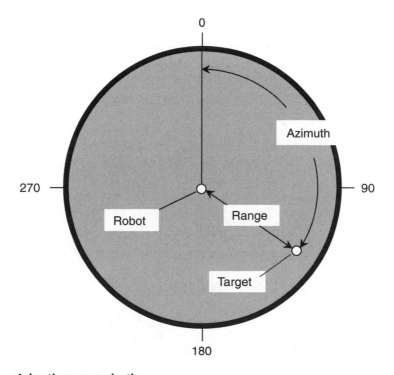

Azimuth-range navigation

A classical *azimuth-range navigation* system is conventional *radar,* which consists of a transmitter, a highly directional antenna, a receiver, and a display. The transmitter produces EM microwave pulses that are propagated in a narrow beam. The EM waves strike objects at various distances. The greater the distance to the target, the longer is the delay before the echo is received. The transmitting antenna is rotated so that all azimuth bearings can be observed.

The basic configuration of an azimuth-range scheme is shown in the illustration. The robot is at the center of the display. *Azimuth* bearings are indicated in degrees clockwise from true north, and are marked around the perimeter. The distance, or *range,* is indicated by the radial displacement.

Some azimuth-range systems can detect changes in the frequencies of returned EM or acoustic pulses resulting from *Doppler effect.* These data are employed to measure the speeds of approaching or receding objects. The robot controller can use this information, along with the position data afforded by the azimuth-range scheme, to navigate in complex environments.

See also **RADAR.**

B

BACK LIGHTING

In a robotic vision system, *back lighting* refers to illumination of objects in the work environment using a light source generally in line with, but more distant than, the objects. The light from the source therefore does not reflect from the surfaces of the objects under observation.

Back lighting is used in situations where the surface details of observed objects are not of interest or significance to the robot, but the shape of the projected image is of importance. Back lighting is also advantageous in certain situations involving translucent or semitransparent objects whose internal structure must be analyzed. Light rays passing through a translucent or semitransparent object can reveal details that *front lighting* or *side lighting* cannot. Compare FRONT LIGHTING and SIDE LIGHTING.

BACK PRESSURE SENSOR

A *back pressure sensor* is a device that detects, and measures, the amount of torque that a robot motor applies at any given time. The sensor produces a signal, usually a variable voltage called the *back voltage,* that increases as the torque increases. The back voltage is used as negative feedback to limit the torque applied by the motor.

When a robot motor operates, it encounters mechanical resistance called back pressure. This resistance depends on various factors, such as the weight of an object being lifted, or the friction of an object as it is moved along a surface. The torque is a direct function of mechanical resistance. As the torque increases, so does the back pressure the motor encounters. Conversely, as the back pressure increases, so does the motor torque necessary to produce a given result.

Back pressure sensors and feedback systems are used to limit the amount of force applied by a robot gripper, arm, drill, hammer, or other device. This can prevent damage to objects being handled by the robot. It also helps to ensure the safety of people working around the robot. The accompanying illustration is a functional block diagram of the operation

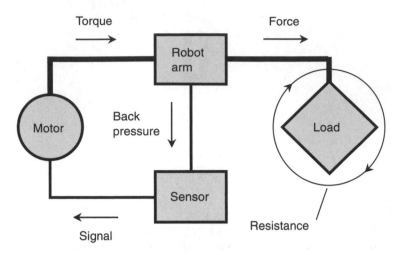

Back pressure sensor

of a back pressure sensor and the associated negative feedback loop that governs the applied torque.

See also ROBOT ARM and ROBOT GRIPPER.

BACKWARD CHAINING

Backward chaining is a logical process that can be used in *artificial intelligence (AI)*. Rather than working with data that have been supplied in advance, the computer requests data as it goes along. In this way, the computer gets only the information it needs to solve a problem. No memory is wasted in storing unnecessary data.

Backward chaining is especially useful in *expert systems,* which are programs designed to help solve specialized problems in unfamiliar fields. A good example is a medical-diagnosis program. Backward chaining can also be of use in electronic troubleshooting, weather forecasting, cost analysis, and even police detective work. Compare FORWARD CHAINING.

See also EXPERT SYSTEM.

BALLISTIC CONTROL

Ballistic control is a form of robotic motion control in which the path, or trajectory, of the device is calculated or programmed entirely in advance. Once the path has been determined, no further corrections are made. The term derives from the similarity to ballistics calculations for aiming guns and missiles.

The main assets of ballistic control are simplicity and moderate cost. A robotic manipulator with ballistic control does not have to carry sensors; a mobile robot with ballistic control does not need its own on-board computer. The main limitation is the fact that ballistic control does not allow for rapid, localized, or unexpected changes in the work environment. Compare CLOSED-LOOP CONTROL.

BANDWIDTH

Bandwidth refers to the amount of frequency space, or *spectrum space,* that a signal requires in order to be transmitted and received clearly. Bandwidth is generally defined as the difference in frequency between the two half-power points in a transmitted or received data signal, as shown in the illustration.

All signals have a finite, nonzero bandwidth. No signal can be transmitted in an infinitely tiny slot of spectrum space. In general, the bandwidth of a signal is proportional to the speed at which the data are sent and received. In digital systems, data speed is denoted in bits per second (bps), kilobits per second (kbps), megabits per second (Mbps), or gigabits per second (Gbps), where

$$1 \text{ kbps} = 1000 \text{ bps} = 10^3 \text{ bps}$$
$$1 \text{ Mbps} = 1000 \text{ kbps} = 10^6 \text{ bps}$$
$$1 \text{ Gbps} = 1000 \text{ Mbps} = 10^9 \text{ bps}$$

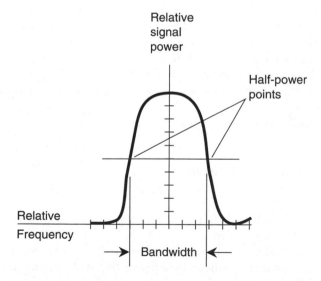

Bandwidth

As the allowable bandwidth is increased, the maximum data speed increases in direct proportion. As the allowable bandwidth is restricted, the maximum data speed decreases in direct proportion.

BAR CODING

Bar coding is a method of labeling objects. Bar-code labels or tags are used extensively in retail stores for pricing and identifying merchandise.

A bar-code tag has a characteristic appearance, with parallel lines of varying width and spacing (see illustration). A laser-equipped device scans the tag, retrieving the identifying data. The reading device does not have to be brought right up to the tag; it can work from some distance away.

Bar coding

Bar-code tags are one method by which objects can be labeled so that a robot can identify them. This greatly simplifies the recognition process. For example, every item in a tool set can be tagged using bar-code stickers, with a unique code for each tool. When a robot's controller tells the machine that it needs a certain tool, the robot can seek out the appropriate tag and carry out the movements according to the program subroutine for that tool. Even if the tool gets misplaced, as long as it is within the robot's work envelope or range of motion, it can easily be found.

See also PASSIVE TRANSPONDER.

BATTERY POWER

See ELECTROCHEMICAL POWER and SOLAR POWER.

BEACON

A *beacon* is a device used to help robots navigate. Beacons can be categorized as either passive or active.

A mirror is a good example of a passive beacon. It does not produce a signal of its own; it merely reflects light beams that strike it. The robot requires a transmitter, such as a flashing lamp or laser beam, and a receiver, such as a photocell. The distance to each mirror can be determined by the time required for the flash to travel to the mirror and return to the robot. Because this delay is an extremely short interval of time, high-speed measuring apparatus is needed.

An example of an active beacon is a radio transmitter. Several transmitters can be put in various places, and their signals synchronized so that they are all exactly in phase. As the robot moves around, the relative phase of the signals varies. Using an internal computer, the robot can determine its position by comparing the phases of the signals from the beacons. With active beacons, the robot does not need a transmitter, but the beacons must have a source of power and be properly aligned.

See also ARTIFICIAL STIMULUS.

BEHAVIOR

In robotics, *behavior* refers to the processing of sensor data into specific motions, sequences of motions, or tasks. There are three main types: *reflexive behavior, reactive behavior,* and *conscious behavior.*

Reflexive behavior is the simplest and fastest form of robotic behavior. Sensors can be, and often are, connected directly to manipulators, propulsion systems, or other mechanical devices. An electric eye that triggers an intrusion alarm is a good example of a device that employs reflexive behavior. When the light beam is broken, an electric current is interrupted, and this actuates an electronic switch that applies power to an acoustic emitter.

Reactive behavior involves a primitive sort of machine intelligence; the extent or nature of the action varies over a range that depends on one or more parameters in the work environment. An example of reactive behavior is the operation of a back pressure sensor, in which the amount of torque applied by a robotic arm or end effector varies depending on the mechanical resistance offered by the manipulated object.

Conscious behavior involves *artificial intelligence (AI)*, in which a robot controller performs complex tasks such as playing chess or making choices that depend on multiple factors that cannot be predicted.

BIASED SEARCH

A *biased search* is an analog method by which a mobile robot can find a destination or target, by first looking off to one side and then "zeroing in."

The illustration shows a biased-search scheme that a boater might use on a foggy day. At some distance from the shoreline, the boater cannot see the dock, but has a reasonably good idea of where it is. Therefore, an approach is deliberately made well off to one side (in this case, to the left) of the dock. When the shore comes into view, the boater turns to the right and follows it until the dock is found.

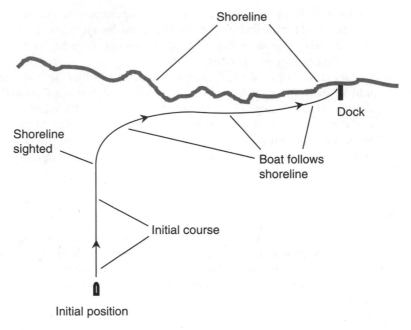

Biased search

For a robot to use this technique effectively, it must have some familiarity with its environment, just as the boater knows roughly where the dock will be. This is accomplished by means of *task-level programming*, a primitive form of *artificial intelligence (AI)*. Compare BINARY SEARCH.

See also TASK-LEVEL PROGRAMMING.

BINARY NUMBER SYSTEM

See NUMERATION.

BINARY SEARCH

In a digital computer, a *binary search,* also called a *dichotomizing search,* is a method of locating an item in a large set of items. Each item in the set is given a number key. The number of keys is always a power of 2. Therefore, when it is repeatedly divided into halves, the end result is always a single key.

If there are 16 items in a list, for example, they might be numbered 1 through 16. If there are 21 items, they can be numbered 1 through 21, with the numbers 22 through 32 as "dummy" keys (unoccupied).

The desired number key is first compared with the highest number in the list. If the desired key is less than half the highest number in the list, then the first half of the list is accepted, and the second half is rejected. If the desired key is larger than half the highest number in the list, then the second half of the list is accepted, and the first half is rejected. The process is repeated, each time selecting half of the list and rejecting the other half, until only one item remains. This item is the desired key.

The illustration shows an example of a binary search to choose one item from a list of 21. Keys are indicated by filled-in squares, except for the desired key, 21, which is indicated by a shaded circle. "Dummy" keys are shown as open squares. Compare BIASED SEARCH.

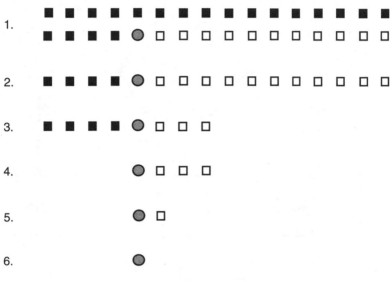

Binary search

BINAURAL MACHINE HEARING

Binaural machine hearing utilizes two sound transducers, spaced a certain minimum distance from each other, to determine the direction from which acoustic waves are coming. This is done by comparing the relative phase and/or the relative loudness of the incoming wavefronts at the transducers.

The human ear/brain system processes acoustic information to a high degree of exactitude, allowing a person to locate a sound source with remarkable accuracy even when the source cannot be seen. When equipped with sensitive transducers, a circuit called a *phase comparator,* and a sophisticated controller, a robot can do the same.

In binaural machine hearing, two sound transducers are positioned on either side of a robot's "head." The phase comparator measures the relative phase and intensity of the signals from the two transducers. These data are sent to the controller, letting the robot determine, with

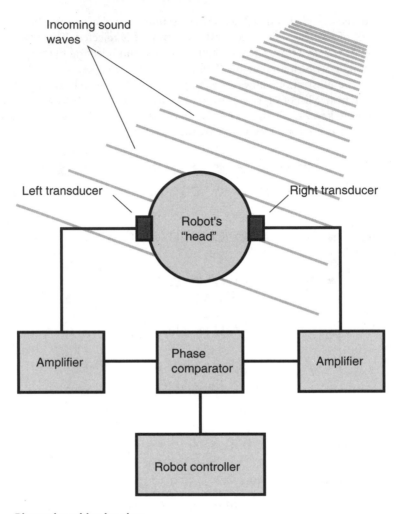

Binaural machine hearing

certain limitations, the direction from which the sound is coming (see the illustration). If the system is confused, the robot head can turn until the confusion is eliminated and a meaningful bearing is obtained.

BINOCULAR MACHINE VISION

Binocular machine vision is the analog of binocular human vision. It is sometimes called *stereoscopic vision*.

In humans, binocular vision allows perception of depth. With one eye, that is, with monocular vision, a human can infer depth to some extent on the basis of perspective. Almost everyone, however, has had the experience of being fooled when looking at a scene with one eye covered or blocked. A nearby pole and a distant tower might seem to be near each other, when in fact they are hundreds of meters apart.

In a robot, binocular vision requires a sophisticated microprocessor. The inferences that humans make, based on what the two eyes see, are exceedingly complicated. The illustration shows the basic concept for binocular machine vision.

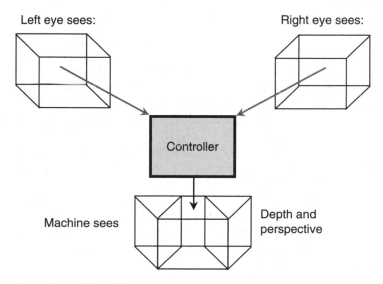

Left eye sees:

Right eye sees:

Controller

Machine sees

Depth and perspective

Binocular machine vision

Of primary importance for good binocular robot vision are the following:

- High-resolution visual sensors
- A sophisticated robot controller
- Programming in which the robot acts on commands, based on what it sees

See also VISION SYSTEM.

BIN PICKING PROBLEM

A *bin picking problem* is a challenge presented to a robotic vision system in which the machine must choose a specific object from a group of objects. Basic machine vision systems can see only the outlines of objects; depth perception is lacking. As viewed from different angles by such a vision system, the appearance of an object can vary dramatically.

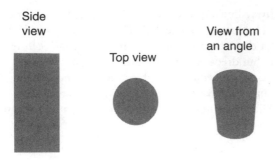

Side
view

Top view

View from
an angle

Bin picking problem

The illustration denotes the example of a cylindrical drinking glass. When seen from exactly side-on, it looks like a rectangle and its interior (left). From the top or the bottom, it looks like a circle and its interior (center). From an intermediate angle, it has a shape similar to that shown at right.

The problem of object recognition is compounded when a certain object must be picked from a bin containing many other objects. Some, most, or all of the desired object can be obscured by other objects. One of the greatest challenges in developing *artificial intelligence (AI)* is giving robots the ability to solve these kinds of problems.

One way to help a robot select items from a bin is to give each item a code. This can be done by means of bar coding or passive transponders.

See also **BAR CODING** and **PASSIVE TRANSPONDER**.

BIOCHIP

A *biochip* is an *integrated circuit (IC)* fabricated with, or from, living matter by means of biological processes. The term has also been suggested for ICs manufactured using techniques that mimic the way nature puts atoms together.

It has been suggested that the human brain is actually nothing more than a sophisticated computer. Any digital computer, no matter how complex, is always built up from individual logic gates. Whether the same can be said for the human brain remains to be seen.

Nature assembles a brain (or any other living matter) by putting protons, neutrons, and electrons together in specific, predetermined patterns. Every proton is identical to every other proton; the same is true for neutrons and electrons. The building blocks are simple. It is the way they are combined that is complicated.

Based on these premises, it is reasonable to suppose that a computer can be "grown" that is as smart as a human brain. Some researchers look at the way nature builds things to get ideas for the construction of improved ICs. The ultimate goal is a biochip that sprouts and evolves as a plant, from a specially engineered "seed."

See also INTEGRATED CIRCUIT.

BIOLOGICAL ROBOT

A *biological robot* is a hypothetical machine derived by cloning from living organisms, and grown in a laboratory environment to perform a specific function or set of functions. Research has been done in this field, although true biological robots have not yet been fabricated or grown.

Biological robots have served as characters in science-fiction stories. The possibilities posed by this notion are limited only by the imagination. There are ethical questions and problems in biological-robot research. These issues are of such grave concern that some scientists refuse to work in this field.

See also CYBORG.

BIOMECHANISM

A *biomechanism* is a mechanical device that simulates the workings of some part of a living body. Examples of biomechanisms are mechanical hands, arms, and legs, known in the medical field as *prostheses*. Especially, the term applies to robotic devices that not only perform the functions of their living counterparts, but look like them.

The term biomechanism can also be used in reference to some body function. Thus, one might talk about the structure of a forearm and hand, calling it a biomechanism. The human anatomy has, in fact, proven to be an excellent model for the design of robotic devices.

See also BIOMECHATRONICS.

BIOMECHATRONICS

The word *biomechatronics* is a contraction of the words biology, mechanics, and electronics. The field of biomechatronics is part of the larger realms of robotics and *artificial intelligence (AI)*. Specifically, biomechatronics involves electronic and mechanical devices that duplicate human body parts and their functions.

Biomechatronics has received more attention in Japan than in the United States. In Japan, some robot researchers attack their problems with religious zeal. Not only would Japanese robotics engineers like to build robots that can do all the things people can do, but some want their robots to look like people, too. The ultimate biomechatronic device is an *android*. Scientists generally agree that an intelligent android will not be developed for many years.

The problem of making androids can be approached from two directions. On the one hand, biological robots might be grown in laboratories by a process of cloning. This idea is clouded by profound ethical issues. On the other hand, engineers can try to build a mechanical robot with the dexterity and intelligence of a human being. This notion, too, brings up ethical questions, but to a lesser degree.

See also ANDROID, BIOCHIP, BIOLOGICAL ROBOT, BIOMECHANISM, and CYBORG.

BIPED ROBOT

A *biped robot* is a robot with two legs that are used for support and propulsion. Usually, but not always, such robots have arms and a head, so they are *androids*.

Physically, biped robots are unstable unless equipped with specialized balancing systems. Humans can manage with two legs because the brain and inner ear together constitute a feedback system that provides a good sense of balance. The human sense of balance can be duplicated electro-mechanically, but the designs are sophisticated and expensive.

Robots that use legs for propulsion generally have four or six legs, because these designs offer better inherent stability than the biped scheme.

See also INSECT ROBOT, QUADRUPED ROBOT, and ROBOT LEG.

BIT-MAPPED GRAPHICS

In a robotic *vision system,* an image can be assembled from thousands of tiny square elements. The smaller the elements, called *pixels,* the more detail the image can show for a given image size. Images made this way are *bit-mapped graphics,* also known as *raster graphics.*

On a computer display, the image you see is a pattern of pixels in a fine, interwoven mesh. You can observe these pixels if you dim your monitor so you can hardly see the image (this is important!) and then look closely at it through a high-powered magnifying lens. A computer stores bit-mapped graphic images as a vast array of logic highs and lows (ones and zeros). To obtain an image from this array of bits, the computer employs a function called a *bit map.*

Bit-mapped graphics always produce approximations of scenes or objects. This is because each pixel is a square, and can take only certain digital values. If the number of pixels in an image is extremely large, the approximation is a good representation of reality in most instances. However, the detail obtainable with bit-mapped graphics is always limited by the *image resolution.*

Bit-mapped graphics produce an artifact called jaggies or aliasing, a peculiar "digitized" appearance in the edges of rendered objects. Vertical and horizontal lines look all right, but curves and diagonals are roughened with "saw teeth." To some extent this can be reduced by means of antialiasing software or photocopy reduction, but a better way is to use *object-oriented graphics.* Compare OBJECT-ORIENTED GRAPHICS.

See also COMPUTER MAP and VISION SYSTEM.

BLACKBOARD SYSTEM

A *blackboard system* incorporates *artificial intelligence (AI)* to help a computer recognize sounds or images. The incoming signal is digitized using an *analog-to-digital converter (ADC).* The digital data is input to a *read/write memory* circuit called the *blackboard.* Then the digital data is evaluated by various specialty programs. The overall scheme is depicted in the diagram.

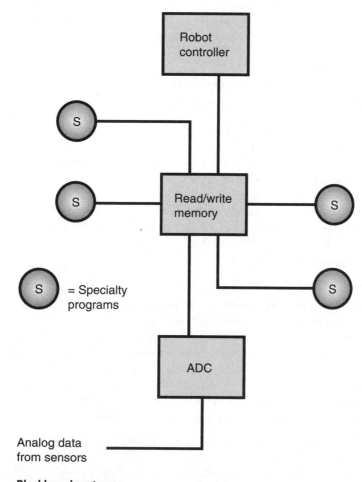

Blackboard system

For speech recognition, specialties include vowel sounds, consonant sounds, grammar, syntax, context, and other variables. For example, a context specialty program might determine whether a speaker means to say "weigh" or "way," or "two," "too," or "to." Another program lets the controller know when a sentence is finished and the next sentence is to begin. Another program can tell the difference between a statement and a question. Using the blackboard as their forum, the specialty circuits "debate" the most likely and logical interpretations of what is heard or seen. A "referee" called a *focus specialist* mediates.

For object recognition, specialties might be shape, color, size, texture, height, width, depth, and other visual cues. How does a computer know if an object is a cup on a table, or a water tower a mile away? Is that a bright lamp, or is it the sun? Is that biped thing a robot, a mannequin, or a person? As with speech recognition, the blackboard serves as a debating ground.

See also OBJECT RECOGNITION and SPEECH RECOGNITION.

BLADDER GRIPPER

A *bladder gripper* or *bladder hand* is a specialized robotic *end effector* that can be used to grasp, pick up, and move rod-shaped or cylindrical objects. The main element of the gripper is an inflatable, donut-shaped or cylindrical sleeve that resembles the cuff commonly used in blood pressure measuring apparatus. The sleeve is positioned so it surrounds the object to be gripped, and then the sleeve is inflated until it is tight enough to accomplish the desired task. The pressure exerted by the sleeve can be measured and regulated using force sensors.

Bladder grippers are useful in handling fragile objects. However, they do not operate fast, and they can function only with objects within a rather narrow range of physical sizes.

See also ROBOT GRIPPER.

BONGARD PROBLEM

The *Bongard problem,* named after its inventor, is a method of evaluating how well a robotic vision system can differentiate among patterns. Solving such problems requires a certain level of *artificial intelligence (AI).*

An example of a Bongard problem is shown in the illustration. There are two groups of six boxes. The contents of the boxes on the left all have something in common; those on the right have the same characteristic in common, but to a different degree, or in a different way. To solve the problem, the vision system (or you) must answer three questions:

- What do the contents of the boxes to the left of the heavy, vertical line have in common?
- What do the contents of the boxes to the right of the line have in common?

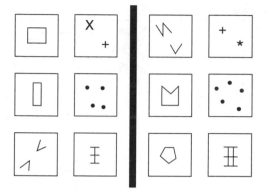

Bongard problem

- What is the difference between the contents of the boxes on opposite sides of the heavy, vertical line?

In this case, the boxes on the left contain four dots or straight lines each; those on the right contain five dots or straight lines each. The difference between the boxes on the left and those on the right, therefore, is in the number of dots or straight lines they contain.

See also OBJECT RECOGNITION.

BOOLEAN ALGEBRA

Boolean algebra is a system of mathematical logic using the numbers 0 and 1 with the operations AND (multiplication), OR (addition), and NOT (negation). Combinations of these operations are NAND (NOT AND) and NOR (NOT OR). Boolean functions are used in the design of digital logic circuits.

In Boolean algebra, X AND Y is written XY or X*Y. NOT X is written with a line or tilde over the quantity, or as a minus sign followed by the quantity. X OR Y is written X+Y. The first table shows the values of these functions, where 0 indicates "falsity" and 1 indicates "truth." The statements

Boolean algebra: basic operations

X	Y	−X	X*Y	X+Y
0	0	1	0	0
0	1	1	0	1
1	0	0	0	1
1	1	0	1	1

Boolean algebra: theorems

Equation	Name (if applicable)
$X + 0 = X$	OR identity
$X * 1 = X$	AND identity
$X + 1 = 1$	
$X * 0 = 0$	
$X + X = X$	
$X * X = X$	
$-(-X) = X$	Double negation
$X + (-X) = X$	
$X * (-X) = 0$	Contradiction
$X + Y = Y + X$	Commutativity of OR
$X * Y = Y * X$	Commutativity of AND
$X + (X * Y) = X$	
$X * (-Y) + Y = X + Y$	
$X + Y + Z = (X + Y) + Z = X + (Y + Z)$	Associativity of OR
$X * Y * Z = (X * Y) * Z = X * (Y * Z)$	Associativity of AND
$X * (Y + Z) = (X * Y) + (X * Z)$	Distributivity
$-(X + Y) = (-X) * (-Y)$	DeMorgan's theorem
$-(X * Y) = (-X) + (-Y)$	DeMorgan's theorem

on either side of the equal sign are logically equivalent. The second table shows several logic equations. These are facts, or theorems. Boolean theorems can be used to analyze complicated logic functions.

See also **LOGIC GATE**.

BRANCHING

Branching refers to routines, or programs, that have points at which an intelligent robot controller must select among alternatives.

Consider a robot on an assembly line that makes cars. The robot's job is to insert hubcaps in the two right-side wheels. (An identical robot does the same job on the left side.) Suppose that 20 percent of the cars are fitted with gold-colored (G) hubcaps; the rest are fitted with silver-colored (S) ones. The robot should insert hubcaps in the following sequence: SS SS SS SS GG SS SS SS SS GG SS SS..., and so on. Every fifth pair of hubcaps is gold.

Each time a hubcap pair is to be inserted, the computer must make a choice. Thus, the routine is at a *branch point* for every hubcap pair. Every fifth time the choice must be made, the robot controller chooses gold hubcaps. Otherwise, it chooses silver ones. This sequence is programmed into the controller. The logical process proceeds something like the flowchart in the accompanying figure.

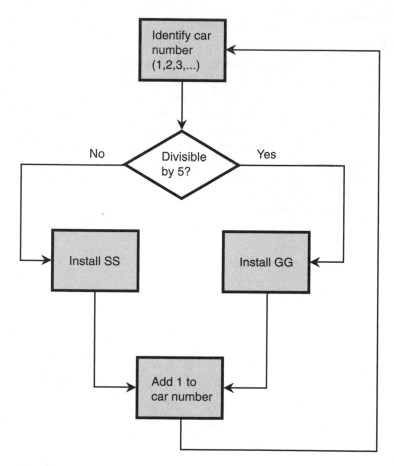

Branching

Suppose a glitch occurs, in which the robot controller or hardware omits or overlooks a single hubcap. This will throw off the robot's perception of the sequence of cars, so it thinks a new car has arrived with each set of rear wheels. Shortly, the front wheel of a car will get a silver hubcap and the rear wheel of the same car will get a gold one. The next car will get a gold hubcap on the front wheel and a silver one on the rear wheel. The repercussions will be repeated down the line over and over, messing up two out of every five cars, or 40 percent of the automobiles coming off the assembly line.

See also EXPERT SYSTEM.

BUMPER

See PROXIMITY SENSING.

BURN-IN

Before any electronic or electromechanical system is put to use, it should undergo a *burn-in* process. This usually involves running the system continuously for hours, days, or weeks. In some cases, a faulty system fails shortly after it is put online. In many instances, however, failure does not occur until a considerable time has passed. Intermittent failures might not manifest themselves until many hours have passed with continuous supervision.

The burn-in process can weed out systems with early-failure problems, minimizing real-time failures.

See also QUALITY ASSURANCE AND CONTROL.

C

CABLE DRIVE

A *cable drive* is a method of transferring mechanical energy in a robotic system from an actuator to a manipulator or end effector. This type of drive can also be used in wheel-drive propulsion systems and in certain indicating devices. The system consists of a cable, or cord, and a set of pulleys.

The main asset of the cable drive is its simplicity. The principal limitation is the fact that the cable can slip on the wheels or pulleys, and over time, the cable can degenerate, and ultimately break without warning. Anyone who has been stranded on a highway because of a failed automotive fan belt can attest to the problems this can cause. Compare CHAIN DRIVE.

CAPACITIVE PRESSURE SENSOR

A *capacitive pressure sensor* consists of two metal plates separated by a layer of nonconductive (dielectric) foam. The resulting variable capacitor is connected in parallel with an inductor; the inductance/capacitance (LC) circuit determines the frequency of an oscillator. If an object strikes the sensor, the plate spacing momentarily decreases. This increases the capacitance, causing a drop in the oscillator frequency. When the object moves away from the transducer, the foam springs back, the plates return to their original spacing, and the oscillator frequency returns to normal. The illustration is a functional block diagram of a capacitive pressure sensor.

The output of the sensor can be converted to digital data using an *analog-to-digital converter (ADC)* and then sent to a robot controller. Pressure sensors can be mounted in various places on a mobile robot, such as the front, back, and sides. Then, for example, physical pressure on the sensor in the front of the robot might send a signal to the controller, which tells the machine to move backward.

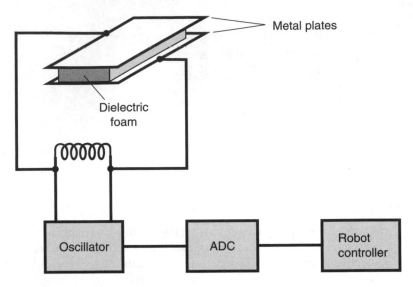

Metal plates

Dielectric
foam

Oscillator

ADC

Robot
controller

Capacitive pressure sensor

A capacitive pressure sensor can be fooled by massive conducting or semiconducting objects in its vicinity. If such a mass comes near the transducer, the capacitance changes, even if direct contact is not made. This phenomenon is known as *body capacitance.* When the effect must be avoided, an *elastomer* can be used for pressure sensing. For *proximity sensing,* however, the phenomenon can be useful.

See also CAPACITIVE PROXIMITY SENSOR, ELASTOMER, and PRESSURE SENSING.

CAPACITIVE PROXIMITY SENSOR

A *capacitive proximity sensor* takes advantage of the *mutual capacitance* that occurs between or among objects near each other.

A capacitive proximity sensor uses a radio-frequency (RF) oscillator, a frequency detector, and a metal plate connected into the oscillator circuit, as shown in the diagram. The oscillator is designed so a change in the capacitance of the plate, with respect to the environment, causes the frequency to change. This change is sensed by the frequency detector, which sends a signal to the apparatus that controls the robot. In this way, if the system is properly designed, a robot can avoid bumping into things. In some detectors, the induced capacitance causes the oscillation to stop altogether.

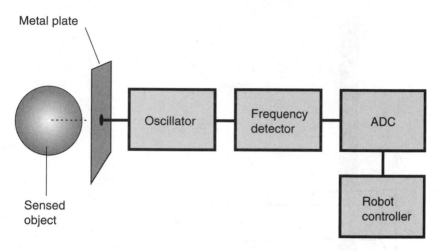

Metal plate

Sensed
object

Capacitive proximity sensor

Objects that conduct electricity to some extent, such as house wiring, people, cars, or refrigerators, are sensed more easily by capacitive transducers than are things that do not conduct, such as wooden chairs and doors. Therefore, other kinds of proximity sensors are necessary for a robot to navigate well in a complex environment, such as a household or office. Compare INDUCTIVE PROXIMITY SENSOR.

See also PROXIMITY SENSING.

CARTESIAN COORDINATE GEOMETRY

Cartesian coordinate geometry is a common method by which a robot manipulator (arm) can move. This term derives from the Cartesian, or rectangular, coordinate system that is used for graphing mathematical functions. Alternatively, this movement scheme is called *rectangular coordinate geometry.*

The drawing shows a Cartesian coordinate system in two dimensions. The axes are perpendicular to each other. In this case, they are up/down (vertical) and left/right (horizontal). Three-dimensional (3-D) Cartesian systems also exist. In a 3-D system, there are three linear axes, with each axis perpendicular to the other two. The manipulator shown in the illustration could be converted to 3-D Cartesian coordinate geometry by allowing the vertical rod to slide forward and backward (in and out of the page) along a horizontal track. Compare CYLINDRICAL COORDINATE GEOMETRY, POLAR COORDINATE GEOMETRY, REVOLUTE GEOMETRY, and SPHERICAL COORDINATE GEOMETRY.

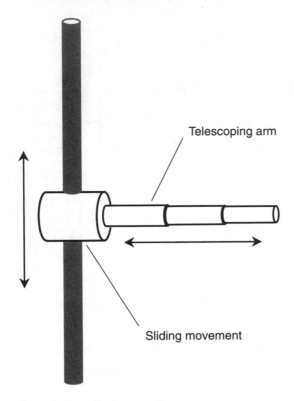

Cartesian coordinate geometry

CENTRALIZED CONTROL

In a system containing more than one robot, *centralized control* refers to oversight of all the individual robots by a single controller. Communication between the controller and the robots is usually done by wireless means such as radio, although other means, such as flexible wire or fiber-optic cables, can be used. This type of robotic system is somewhat analogous to a client-server computer network.

In a centrally controlled robotic system, the main computer plays the role of a quasi-human operator. In some systems, the individual robots are partially autonomous, containing controllers of their own; this allows the system to keep operating at full capacity for a time, even in the event of a break in one or more of the communication links. This is known as *partially centralized control*. Another example of partially centralized control is a system in which each robot receives a set of instructions from the controller, stores those instructions, and then carries them out independently of the central controller.

In some robotic systems, the individual units are completely and continuously dependent on the central controller, and cannot function if the communication link is severed. Such a system is said to employ *fully centralized control*. Compare DISTRIBUTED CONTROL.

See also AUTONOMOUS ROBOT and INSECT ROBOT.

CHAIN DRIVE

A *chain drive* is a method of transferring mechanical energy in a robotic system from an actuator to a manipulator or *end effector*. It can also be used in wheel-drive propulsion systems. The system consists of a chain and a set of wheels with sprockets.

The main asset of the chain drive is its simplicity. It can provide additional traction compared with a cable drive, because the chain is not likely to slip on the sprockets. Another asset is the fact that variable speed and power can be obtained by using sprockets of various sizes, in conjunction with a shifting mechanism. On the downside, the chain can come off the sprockets. The chain requires lubrication and maintenance, and can be noisy in operation. A common example of a chain drive is found in any bicycle. Compare CABLE DRIVE.

CHARGE-COUPLED DEVICE (CCD)

A *charge-coupled device (CCD)* is a camera that converts visible-light images into digital signals. Some CCDs also work with infrared (IR) or ultraviolet (UV). Common digital cameras work on a principle similar to that of the CCD.

The image focused on the retina of the human eye, or on the film of a conventional camera, is an *analog image*. It can have infinitely many configurations, and infinitely many variations in hue, brightness, contrast, and saturation. A digital computer, however, needs a *digital image* to make sense of, and enhance, what it "sees." Binary digital signals have only two possible states: high and low, or 1 and 0. It is possible to get an excellent approximation of an analog image in the form of high and low digital signals. This allows a computer program to process the image, bringing out details and features that would otherwise be impossible to detect.

The illustration is a simplified block diagram of a CCD. The image falls on a matrix containing thousands or millions of tiny sensors. Each sensor produces one *pixel* (picture element). The computer (not shown) can employ all the tricks characteristic of any good graphics program. In

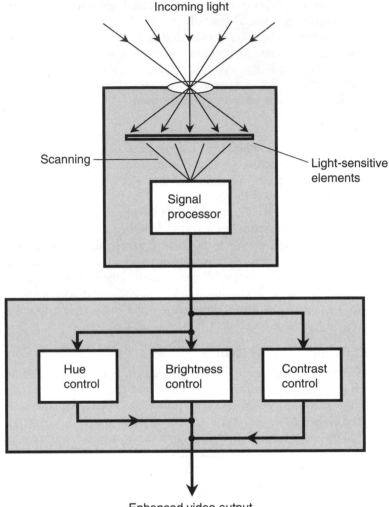

Incoming light

Scanning

Light-sensitive elements

Signal processor

Hue control

Brightness control

Contrast control

Enhanced video output

Charge-coupled device

addition to rendering high-contrast or false-color images, the CCD and computer together can detect and resolve images much fainter than is possible with conventional camera film or more primitive types of video cameras. This makes the CCD useful in robots that must employ night vision. Compare IMAGE ORTHICON and VIDICON.

See also VISION SYSTEM.

CHECKERS AND CHESS

A computer can be programmed to play *checkers*. An excellent program was created by the roboticist Arthur Samuel, in which the computer can not only play the game move by move, but can also look ahead, or anticipate, to see the possible consequences of a move.

Checkers is a fairly simple board game. It is more complex than tic-tac-toe, but far less sophisticated than chess. Anyone who has played tic-tac-toe has discovered that it is always possible to get at least a draw (tie). This is so elementary that a high-school student with some programming experience can get a computer to play tic-tac-toe. In this game the machine needs to look only one move ahead.

Look-ahead strategy involving more than one move takes a certain amount of practice or learning. Computers can, however, be programmed to learn from their mistakes. Arthur Samuel's checkers program uses a multiple-move look-ahead strategy so effectively that even the best human players in the world find it nigh impossible to beat his machine.

There is another scheme that can be used for checkers: adopt a general game plan. General strategies can be broadly categorized as either defensive or offensive. The defensive/offensive schemes require look-ahead of only one move.

Chess has been used to develop and test machine intelligence. One of the first chess-playing machines was developed by Rand Corporation in 1956. Chess is a complex game. A computer must look ahead more than one move to play a good game of chess. Multiple look-ahead strategy, along with machine learning, can enable a computer to play chess at a level of skill comparable to the masters.

The program developed by Rand Corporation was able to prove some mathematical theorems. This is another good way to test the intelligence of a computer.

CLEAN ROOM

A *clean room* is a chamber specially designed and operated to minimize airborne contaminants. In some industries it is important that dust, dirt, bacteria, and other particulates be kept to an absolute minimum. A good example is the manufacture of *integrated circuits (ICs)* for electronic and computer systems. Robots have a considerable advantage over people in these environments.

If certain precautions are observed, the environment in a room can be kept "clean" while allowing humans in. People who enter such a room must first put on airtight suits, gloves, and boots. A room that only robots enter, not people, can always be just a little bit cleaner.

The contamination in a clean room is measured in terms of the number of particles of a certain size in 1 liter (1000 cubic centimeters) of air. Alternatively, the cubic foot is used as the standard unit of volume.

See also INTEGRATED CIRCUIT.

CLINOMETER

A *clinometer* is a device for measuring the steepness of a sloping surface. Mobile robots use clinometers to avoid inclines that might cause them to tip too far, possibly even falling over.

The floor in a building is almost always horizontal. Thus, its incline is zero. But sometimes there are inclines such as ramps. A good example is the kind of ramp used for wheelchairs, in which a very small elevation change occurs. A rolling robot cannot climb stairs, but it might use a wheelchair ramp, provided the ramp is not so steep that it upsets the robot's balance or causes it to spill or drop its payload.

A clinometer produces an electrical signal whenever it is tipped. The greater the angle of incline, the greater is the electrical output, as shown on the left side of the graph. A clinometer might also show whether an incline goes down or up. A downward slope might cause a negative voltage at the transducer output, and an upward slope a positive voltage, as shown on the right side of the graph.

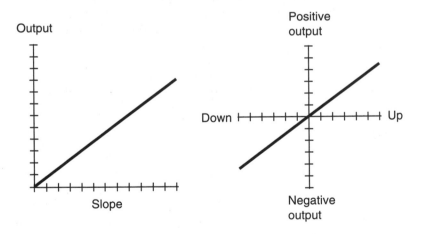

Clinometer

CLOSED-LOOP CONTROL

Closed-loop control is a form of robot manipulator motion control in which the path, or trajectory, of the device is corrected at frequent intervals.

After motion begins, a position sensor detects possible errors in the trajectory. If an error is detected, the sensor outputs a signal that operates through a *feedback* circuit to bring the manipulator back on course. The term derives from the fact that the feedback and control-signal circuits together constitute a closed loop.

The main asset of closed-loop control is accuracy. In addition, closed-loop control can compensate for rapid, localized, or unexpected changes in the work environment. The principal disadvantages are greater cost and complexity than simpler schemes such as *ballistic control.* Compare BALLISTIC CONTROL.

CLOSED-LOOP SYSTEM

A *closed-loop system* is a set of devices that regulates its own behavior. Closed loops can be found in many kinds of machines, from the engine in a car (governor) to the gain control in a radio receiver (automatic level control).

A closed-loop system, also known as a *servomechanism,* has some means of incorporating mechanical *feedback* from the output to the input. A sensor at the output end generates a signal that is sent back to the input to regulate the machine behavior. A good example of this is a *back pressure sensor.* Another example is *closed-loop control* of a robot manipulator. Compare OPEN-LOOP SYSTEM.

See also BACK PRESSURE SENSOR, CLOSED-LOOP CONTROL, and SERVOMECHANISM.

CMOS

See COMPLEMENTARY METAL-OXIDE SEMICONDUCTOR.

COEXISTENCE

The term *coexistence* refers to programmed interactions among *insect robots* that share a working environment. The robots in such a system do not communicate directly with each other, but they all communicate with a central controller. There are three general schemes: *ignorant coexistence, informed coexistence,* and *intelligent coexistence.*

In ignorant coexistence, none of the robots is aware that any of the others exists. In this sense, when two robots encounter one another, each machine regards its counterpart as an obstruction. Most mobile robots are programmed to avoid obstacles and hazards, maintaining a minimum distance of, say, 1 m. Thus, if there are numerous robots in a given environment and they all have ignorant coexistence, they tend to stay away from each other. If the robot "population density" is moderate to high, the machines tend to be more or less evenly spaced in the work environment at all times.

In informed coexistence, mobile robots can differentiate between obstructions or hazards and other robots. In this type of system, the robots are programmed to react or behave in a specific, but simple, way toward their counterparts. The most common behavior is for a robot to execute a specific set of movements when it senses the proximity of another robot, and a different set of movements when it senses the proximity of a nonrobotic obstruction or hazard. An example is for the machine to stop and reverse direction if it comes near an obstruction; but if it comes near another robot, it stops, waits a second, and if the other robot remains in the way, turns right 90°, proceeds 1 m, then turns left 90° and resumes moving in the original direction.

In intelligent coexistence, as in informed coexistence, the robots can differentiate between obstructions or hazards and other robots. However, the programmed response is more sophisticated. For example, each robot might be programmed to avoid coming within 1 m of any other robot. If such an approach does occur, triggering the avoidance response, the robot is programmed to move in a direction corresponding to the average direction of all the other robots in the system. Each robot obtains this general information from the controller. Compare COOPERATION.

See also AUTONOMOUS ROBOT, CENTRALIZED CONTROL, DISTRIBUTED CONTROL, and INSECT ROBOT.

COGNITIVE FATIGUE

Cognitive fatigue is a form of mental exhaustion sometimes experienced by users of *telepresence* systems. Most teleoperated systems must compromise realism in order to keep within limitations imposed by available bandwidth and allowable expense.

In a typical telepresence system, the cameras usually lack peripheral vision. Signal propagation delays can cause latency problems (time lag between command and response), particularly when teleoperation is done over long distances. Image resolution (detail) and refresh rate (the number of video frames per second) are generally compromised. Audio systems are generally better than video systems because the necessary bandwidths are smaller, but tactile sensation is poor or absent.

Symptoms of cognitive fatigue include wandering attention, sleepiness, headache, and irritability. These problems can result in equipment operation errors.

See also TELEPRESENCE.

COGNIZANT FAILURE

Cognizant failure is a feature of machine intelligence in which a failed subsystem or program is replaced by one at a higher level, while ensuring that all processes continue to run smoothly without undesired side effects.

In uncomplicated systems, a high-level part of the system can temporarily take over the tasks of a lower-level part, without regard for details of the event. In scenarios where the possibilities are diverse and variable, certain additional procedural steps, not normally necessary, are sometimes required to ensure smooth operation while the lower-level device or subsystem is repaired.

Consider the case of a *smart home* equipped with smoke detectors, heat sensors, a telephone link to the fire department, and a set of sprinklers. What should the system do if a heat sensor is set off by a mischievous child with a hair dryer, causing a false alarm? An unsophisticated system calls the fire department and actuates the sprinklers, causing embarrassment and unnecessary damage to furniture. A sophisticated system can prevent these undesirable things from taking place, provided the owner of the house, or some backup sensing system, is present to determine that there is actually no fire. The owner or backup system must be cognizant of the fact that the alarm is false. Then the sprinkler system can be disabled, a call can be made to the fire department to cancel the alarm, and the offending sensor, if it has been damaged, can be shut down until it is replaced. (The child can be disciplined as well, although this is the responsibility of the human home owner.) Compare GRACEFUL DEGRADATION.

COLOR SENSING

Many robotic *vision systems* function only in grayscale. Color sensing can be added, in a manner similar to the way it is added to television (TV) systems.

Color sensing can help a robot determine the identity or nature of an object. Is an observed horizontal surface a floor or a grassy yard? (If it is green, it is probably a grassy yard.) Sometimes, objects have regions of different colors that have identical brightness as seen by a grayscale system. Such objects can be analyzed better with a color-sensing system than with a vision system that sees only shades of gray.

The drawing shows a block diagram of a color-sensing system. Three grayscale cameras are used. Each camera has a color filter in its lens. One filter is red, another is green, and another is blue. These are the three *primary colors* of radiant light. All possible hues, brightnesses, and saturations are comprised of these three colors in various ratios. The signals from the three cameras are processed by a microcomputer, and the result is fed to the robot controller.

See also GRAYSCALE, OBJECT RECOGNITION, TEXTURE SENSING, and VISION SYSTEM.

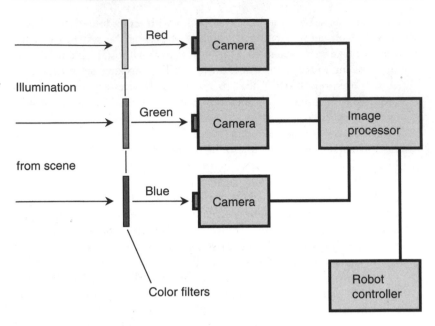

Color sensing

COMPETING SENSORS

See SENSOR COMPETITION.

COMPLEMENTARY METAL-OXIDE SEMICONDUCTOR (CMOS)

Complementary metal-oxide semiconductor, also called *CMOS* (pronounced "seamoss"), is the name for a technology used in digital devices, such as computers. Two types of *field-effect transistor (FET)* work together, in tandem and in huge numbers, on a single *integrated circuit (IC)* chip.

The main asset of CMOS technology in robotics is the fact that the devices can function effectively with tiny electrical currents. Thus, well-engineered CMOS circuits draw very little power from the power supply, allowing the use of batteries. Another advantage of CMOS technology is that it works extremely fast. It can process a lot of data in a short period of time.

A disadvantage of CMOS devices is the fact that they are easily damaged by static electricity. Devices of this type must be stored with their pins embedded in conductive foam material, and/or packaged in special plastic that resists electrostatic-charge buildup. When constructing or servicing equipment using CMOS, technicians must take precautions to avoid the presence of static electric charges on their hands, and on instruments such as probes and soldering irons. This is usually ensured by physically connecting the technician's body to a good electrical ground.

See also INTEGRATED CIRCUIT.

COMPLEX-MOTION PROGRAMMING

As machines become smarter, the programming gets more sophisticated. No machine has yet been built that has intelligence anywhere near that of a human being. Some researchers think that true *artificial intelligence (AI)*, at a level near that of the human brain, will never be achieved.

The programming of robots can be divided into levels, starting with the least sophisticated and progressing to the theoretical level of true AI. The drawing shows a four-level scheme. Level 2, just below the task level but above the simple-motion level, is called *complex-motion programming*. Robots at this level can perform sets of motions in defined sequences. Compare ARTIFICIAL INTELLIGENCE, SIMPLE-MOTION PROGRAMMING, and TASK-LEVEL PROGRAMMING.

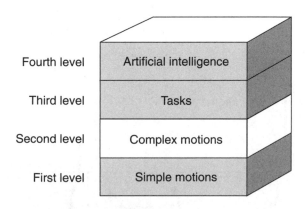

Fourth level	Artificial intelligence
Third level	Tasks
Second level	Complex motions
First level	Simple motions

Complex-motion programming

COMPLIANCE

Compliance is the extent to which a robot *end effector* or manipulator moves, or yields, when a force is applied to it. It can be expressed qualitatively (using terms such as "springy" or "rigid") or quantitatively in terms of displacement per unit force (such as millimeters per newton).

A robot is said to be compliant if its mechanical movements are affected by external forces, including linear pressure or torque. The compliance can occur along one, two, or three axes, or in a rotational sense. Generally, a *compliant robot* should be adjusted so the behavior of its

manipulators and end effectors keeps the stress on its components to a minimum. One means of accomplishing this is a *back pressure sensor*.
See also BACK PRESSURE SENSOR.

COMPOSITE VIDEO SIGNAL

A *composite video signal* is the waveform that modulates a television (TV) or video carrier. The composite signal contains video intelligence as well as synchronization, blanking, and timing pulses. The bandwidth is typically 6 MHz (6 megahertz) for conventional fast-scan signals, and approximately 3 kHz (3 kilohertz) for slow-scan signals. A video camera, such as an image orthicon or vidicon, produces a fast-scan signal. Some robotic vision systems generate and analyze composite video signals.

The illustration shows the waveform for a single line of a *color picture signal*. There are normally 525 or 625 lines in a complete frame for standard

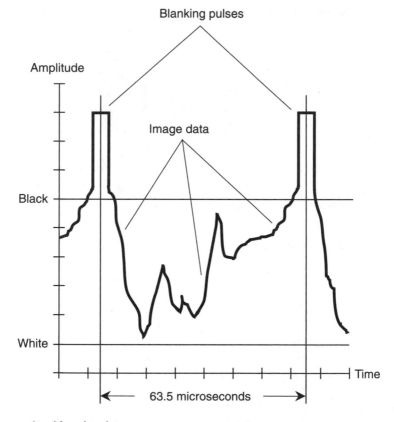

Composite video signal

fast-scan video. In robotic vision systems, there are advantages to using more lines per frame than is standard with television, in order to obtain improved image resolution.

See also IMAGE ORTHICON, VIDICON, and VISION SYSTEM.

COMPUTER MAP

An autonomous robot must have a sense of where it is relative to surrounding objects, so that it will not bump into things, and so that it can find whatever it is seeking. For this to be possible, the robot controller can make a *computer map* of its environment.

Computer maps can be generated using *radar, sonar,* or a *vision system.* Such a map can exist in either two or three dimensions. A two-dimensional (2-D) computer map of the objects in a room might be generated for a flat plane 1 m above the floor. Several 2-D maps, representing various altitudes above the floor, can be combined to create a composite three-dimensional (3-D) map.

A more sophisticated method of generating a 3-D computer map involves the use of *spherical coordinates.* The spherical coordinate system defines *azimuth* (compass bearing), *elevation* (angle above the horizontal), and *range* (radial distance). For such a map to serve its purpose, hundreds or even thousands of individual soundings or observations must be made. These soundings or observations should be distributed evenly around a half-sphere above the horizontal for terrestrial robots, or around a full sphere for submarine, airborne, or deep-space robots. In deep space, a reference plane must be chosen to serve as the "horizontal." The larger the number of soundings, the better is the resolution of the map.

See also RADAR, SONAR, and VISION SYSTEM.

CONFIGURATION SPACE (C-SPACE)

A *configuration space* (abbreviated *C-space*) is a scheme in which the location and orientation of a robot is determined relative to other objects in its environment. Ideally, a C-space should use the minimum number of coordinates necessary to accomplish this task. This eliminates redundancy, which consumes controller memory and can cause confusion.

Consider a *mobile robot* designed to function on a single floor of a building. The total physical region in which this robot exists (the *world space*) is three-dimensional (3-D). This constitutes three *degrees of freedom,* which can be considered in terms of the Cartesian (rectangular) coordinates x (north/south), y (east/west), and z (up/down). The orientation, or attitude, of the robot, can require up to three additional degrees of freedom: *p (pitch), r (roll),* and *w (yaw).*

On the flat plane of a floor, the location of the robot can be denoted in two-dimensional (2-D) coordinates. In the Cartesian system described above, these are x and y. If the attitude nevertheless requires that p, r, and w each be specified, the C-space requires five degrees of freedom: x, y, p, r, and w. However, p, r, and w may not all be important in the 2-D case. This could reduce the number of degrees of freedom in the C-space still further. Compare WORK ENVELOPE and WORK ENVIRONMENT.

See also CARTESIAN COORDINATE GEOMETRY, PITCH, ROLL, and YAW.

CONTACT SENSOR

A *contact sensor* is a device that detects objects, obstructions, or barriers by means of direct physical contact. Contact sensors can also be used to measure applied force or torque. In robotics, such devices include "whiskers" and pressure sensors.

Simplicity is the main asset of contact sensing when used to determine the presence or absence of an object. In order to measure force or torque accurately, especially when such force or torque must be regulated, a *closed-loop system* is required.

See also CLOSED-LOOP CONTROL, CLOSED-LOOP SYSTEM, FEEDBACK, and PRESSURE SENSING.

CONTEXT

Context is the environment in which a word is used. It is important in *speech recognition* systems, such as those used in personal or security robots designed to respond to spoken commands.

Everyone has heard the expression "out of context." When a word is used out of context, it results in a phrase or sentence that does not make sense. Worse yet, it might mean something not intended. When a word is taken out of context, the phrase or sentence is technically all right, but it is interpreted as nonsense, or in the wrong way.

In order to interpret and respond to spoken statements properly, a computer or robot with artificial intelligence must know the context in which each word is used. Humans have an innate sense of context; machines do not. This makes the design and programming of effective speech recognition systems an extremely sophisticated business.

See also PROSODIC FEATURES and SPEECH RECOGNITION.

CONTINUOUS ASSISTANCE

See SHARED CONTROL.

CONTINUOUS-PATH MOTION

A robot arm can move smoothly or in discrete steps. Smooth-moving robot manipulators employ *continuous-path motion*.

For a robot to move along a smooth, continuous path, every point along the way must, in theory, be stored in the controller memory. Of course, this is not literally possible, because a continuous path contains an infinite number of points. Continuous-path motion uses mathematical *functions,* rather than point sets, to define the instantaneous position of a robot manipulator.

In the function method, the instantaneous position is stored as a set of mathematical functions. Such motion is truly continuous, in that it actually passes through an infinite number of points. This is possible because of the smooth nature of the mathematical functions. This principle is the robot-motion analog of vector graphics in computing. Compare POINT-TO-POINT MOTION.

CONTROLLER

In a robot, the *controller* is a computer that oversees and controls the operation and motion of the machine. The illustration is a functional block diagram of a controller. The heart of the controller is the *central processing unit (CPU),* which is similar to the CPU in a personal computer. Movement instructions are held in *random-access memory (RAM)* and/or on *storage media* such as a hard drive.

The interface does several things. Mainly, it allows the microcomputer to communicate with a human operator or supervisor. Through the interface,

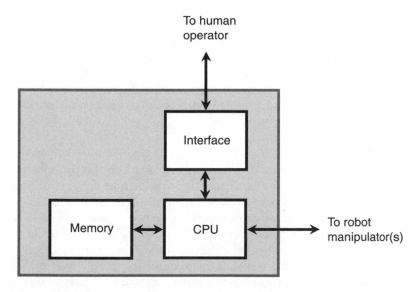

Controller

it is possible to reprogram the memory to change the movement instructions. The actions or function repertoire of the robot can be displayed on a monitor screen. There might also be various malfunction indicators. Some of the more sophisticated interfaces have a *teach box*, which lets the human operator reprogram the motions and path of the robot.

See also TEACH BOX.

CONTROL TRADING

Control trading is a limited form of robotic remote control in a system that employs *teleoperation*. The operator instructs the robot to perform a specific, complete task, such as vacuuming a room or mowing a lawn. The machine then carries out the entire task without further instruction or supervision by the human.

Control trading has obvious assets. The human operator does not have to constantly monitor the progress of the machine, although periodic checking is advisable to ensure that a major malfunction does not occur. It is thus possible for a single operator to oversee the operation of several robots at the same time. Another asset is the fact that *latency*, or the time lag caused by signal propagation delays, is not a serious problem. Control trading is ideal, for example, in the teleoperation of a robot on Mars, or the teleoperation of an interplanetary space probe. Still another asset is that large signal bandwidth is not required, especially for the uplink to the machine; commands can consist of encoded messages of a relatively small number of bytes.

The main limitation of control trading is the fact that the robot cannot be expected to contend with sudden, unforeseen changes in the work environment. The machine performs its programmed set of operations under the assumption that the environment will cooperate. In scenarios where the robot work environment is subject to frequent change, *shared control* is generally superior to control trading. Compare SHARED CONTROL.

See also TELEOPERATION.

COOPERATION

Cooperation is constructive or synergistic interaction of robots in a system. It can take various forms, depending on the manner and extent to which the robots communicate, and the degree of autonomy each machine has.

In *nonactive cooperation*, the robots do not necessarily have to communicate. However, it is important that each robot be able to tell the other robots apart from general objects in the environment. This prevents undesirable conditions such as collisions between robots, multiple robots attempting the same task at the same time and in the same place, and uneven distribution of the machines in the work environment. Other than the ability to avoid conflicting with its peers, each robot in a non-active-cooperative system need not pay particular attention to the others. In a well-designed system of this kind, cooperation occurs naturally.

In *active cooperation,* the robots are capable of acknowledging one another, and in some cases communicating with and assisting each other as well. Active cooperation can range from "loose," in which the machines are aware of each other's existence and function but do not communicate, to "tight," in which each robot can communicate with any or all of the others. Some systems can be engineered to exhibit *cooperative mobility,* in which two or more robots can combine in "special teams" to deal with complex or difficult tasks that a single robot cannot carry out. A special form of active cooperation involves *centralized control,* in which the robots are all dependent on oversight by a single controller. Compare COEXISTENCE.

See also AUTONOMOUS ROBOT, CENTRALIZED CONTROL, DISTRIBUTED CONTROL, and IN-SECT ROBOT.

CORRESPONDENCE

In *binocular machine vision,* the term *correspondence* refers to the focusing of both video cameras or receptors on the same point in space. This ensures that the video perception is correct. If the two "eyes" are not focused at the same point, the ability of the machine to perceive depth is impaired.

The human sense of correspondence can be confused when looking at a grid of dots, or at a piece of quadrille graph paper. The illustration shows

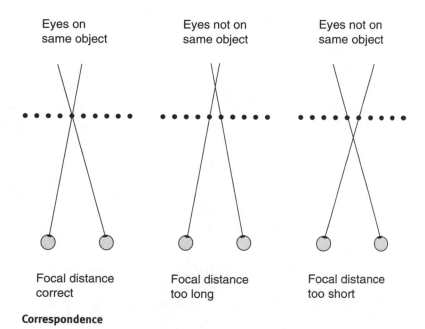

Correspondence

two ways that human eyes, or a machine vision system, can be fooled by such a pattern. This problem is generally limited to observations of regular patterns of dots, squares, or other identical objects. It rarely occurs in complex scenes in which geometric shapes do not repeat.

On the left in the illustration, both video sensors (shown as eyeballs) are looking at the same point. Thus, depth is perceived correctly, even if the views of the object appear slightly different because of the difference in viewing angle through either sensor. In the drawings in the center and on the right, the left sensor is looking at one object in the set, while the right sensor looks at another. Because all the objects are evenly spaced, they seem to line up as perceived by the vision system. If a robot manipulator acts on this incorrect information, positioning errors are likely.

See also BINOCULAR MACHINE VISION.

CRYPTANALYSIS

Cryptanalysis is the art of breaking *ciphers,* which are signal-processing schemes used to keep unauthorized people from intercepting communications or gaining access to sensitive data. With the help of computers, cryptanalysis has become much more sophisticated than it once was. A computer can test different solutions to a code much more rapidly than teams of humans ever could. Beyond that, *artificial intelligence (AI)* can be employed in an attempt to figure out what an enemy is thinking. This streamlines the process of cipher breaking. It lets the *cryptanalyst,* or code breaker, get a feel for the general scheme behind a cipher, and in this way, it helps the cryptanalyst understand the subtleties of the code more quickly.

One of the earliest cryptanalysts to use a computer was *Alan Turing,* known as a pioneer in AI. In the early 1940s, during World War II, the Germans developed a sophisticated machine called *Enigma* that encoded military signals. The machine and its codes confounded Allied cryptanalysts, until Alan Turing designed one of the first true computers to decode the signals.

As computers become more powerful, they can create more complex ciphers. But they can also invent increasingly sophisticated decryption schemes. In warfare, the advantage in encryption/decryption goes to the side with the more advanced AI technology.

CYBERNETICS

The term *cybernetics* refers to the science of goal-seeking, or self-regulating, things. The word itself comes from the Greek word for "governor." The fields of robotics and artificial intelligence are subspecialties within the science of cybernetics. Computer-controlled robots that interact with their environments are cybernetic machines.

An example of a cybernetic process is pouring a cup of coffee. Suppose someone says to a personal robot, "Please bring me a cup of coffee, and be sure it's hot." In the robot controller's memory, there are data concerning what a coffee cup looks like, the route to the kitchen, the shape of the coffee pot, and a relative-temperature-interpretation routine, so the robot knows what the person means by "hot." A personal robot must go through an unbelievably complicated process to get a cup of coffee. This becomes evident when one tries to write down each step in rigorous form.

CYBORG

The word *cyborg* is a contraction of "cybernetic" and "organism." In robotics, the term refers to a human whose body is comprised largely, or even mostly, of robotic elements, but who is still biologically alive.

If a person is given a single robotic hand or arm, it is called a *bionic* body part or *prosthesis*. Science fiction carries this notion to the point that a person seriously injured might be reconstructed significantly, or even almost entirely, of bionic parts. Such a being would be a true cyborg. Technology is a long way from creating cyborgs, but some scientists believe they will someday be common. A few futurists envision a society comprised of human beings, cyborgs, smart robots, and computers. This has been called a *cybot society.*

While enthusiasm for the idea of a cybot society runs high in Japan, there is somewhat less interest in the United States and Europe. Americans and Europeans think of robots as serving mainly industrial purposes, but the Japanese think of them as being in some sense alive. This might be why the Japanese are so much more active in developing human-like robots.

See also ANDROID and PROSTHESIS.

CYLINDRICAL COORDINATE GEOMETRY

Cylindrical coordinate geometry, also known as *cyclic coordinate geometry,* is a scheme for guiding a *robot arm* in three dimensions. A cylindrical coordinate system is a polar system with an extra coordinate added for *elevation.* Using this system, the position of a point can be uniquely determined in three-dimensional (3-D) space.

In the cylindrical system, a reference plane is used. An origin point is chosen, and also a reference axis, running away from the origin in the reference plane. In the reference plane, the position of any point can be specified in terms of the *reach,* or distance from the origin, and the *base rotation,* which is the angle measured counterclockwise from the reference axis.

The elevation coordinate is either positive (above the reference plane), negative (below it), or zero (in it).

The illustration shows a robot arm equipped for cylindrical coordinate geometry. Compare CARTESIAN COORDINATE GEOMETRY, POLAR COORDINATE GEOMETRY, REVOLUTE GEOMETRY, and SPHERICAL COORDINATE GEOMETRY.

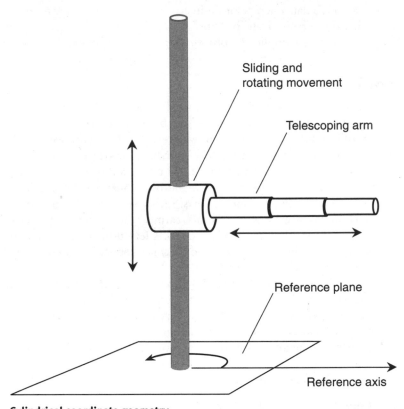

Sliding and rotating movement

Telescoping arm

Reference plane

Reference axis

Cylindrical coordinate geometry

D

DATA COMPRESSION

Data compression is a method of maximizing the amount of digital information that can be stored in a given space, or sent in a certain period of time.

Text and program files can be compressed by replacing often-used words and phrases with symbols such as =, #, &, or @, as long as none of these symbols occurs in the uncompressed file. When the data are received, they are uncompressed by substituting the original words and phrases for the symbols.

Digital images can be compressed in either of two ways. In *lossless image compression,* detail is not sacrificed; only the redundant bits are eliminated. In *lossy image compression,* some detail is lost, although the loss is usually not significant.

Text and programs can generally be reduced in size by about 50 percent by means of data compression. Images can be reduced to a much larger extent if a certain amount of loss can be tolerated. Some advanced image-compression schemes can output a file that is only a tiny fraction of the original file size.

DATA CONVERSION

Many communications systems "digitize" analog signals at the *source* (transmitting end) and "undigitize" the signals at the *destination* (receiving end). Digital data can be transferred bit by bit (serial) or in bunches (parallel). *Data conversion* is the process of altering data between analog and digital forms, or between parallel and serial forms.

Analog to digital

Any analog, or continuously variable, signal can be converted into a string of pulses whose amplitudes have a finite number of states. This is *analog-to-digital (A/D) conversion.*

An *A/D converter* or *ADC* samples the instantaneous amplitude of an analog signal and outputs pulses having discrete levels, as shown in Fig. 1. The number of levels is called the *sampling resolution,* and is usually a power of 2. The number of pulses per second is the *sampling rate.* The time between pulses is the *sampling interval.* In this example, there are eight levels, represented by three-digit binary numbers from 000 to 111.

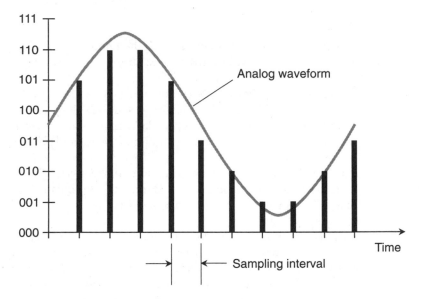

Data conversion—Fig. 1

In general, the minimum workable digital sampling rate is approximately twice the highest analog data frequency. This is a general principle in communications engineering, known as the *Nyquist theorem* or *sampling theorem.* For a signal with components as high as 3 kHz, the minimum sampling rate is 6 kHz. The commercial voice standard is 8 kHz. For hi-fi music transmission, the standard sampling rate is 44.1 kHz. In machine communications systems, the minimum sampling rate depends on the speed with which data must be transferred between points, such as from a central controller to a mobile robot.

Digital to analog

The scheme for *digital-to-analog (D/A) conversion* depends on whether the signal is binary or multilevel. The D/A conversion process is carried out by a *D/A converter (DAC)*.

In a binary DAC, a microprocessor reverses the A/D conversion process done in recording or transmission. Multilevel digital signals can be converted back to analog form by "smoothing out" the pulses. This can be intuitively seen by examining Fig. 1. Imagine the train of pulses being smoothed into the continuous curve.

Digital signals lend themselves to repeated reproduction without loss of integrity. Digital signals are also relatively immune to the effects of noise in wireless and long-distance cable circuits. For this reason, even if the initial input and final output signals are analog in nature, such as moving images or human voices, there are advantages to using digital format in the intervening medium.

Digital signals can be clarified by means of *digital signal processing (DSP)* to enhance the *signal-to-noise (S/N)* ratio, thereby minimizing the number of communication errors and necessary bandwidth while maximizing the data transfer rate. This is true whether the ultimate input and output signals are analog or digital.

Serial versus parallel

Binary data can be sent and received one bit at a time along a single line or channel. This is *serial data transmission.* Higher data speeds can be obtained by using multiple lines or a wideband channel, sending independent sequences of bits (high and low, or 1 and 0) along each line or subchannel. This is *parallel data transmission.*

In *parallel-to-serial (P/S) conversion,* bits are received from multiple lines or channels, and transmitted one at a time along a single line or channel. A *buffer* stores the bits from the parallel lines or channels while they are awaiting transmission along the serial line or channel. In *serial-to-parallel (S/P) conversion,* bits are received from a serial line or channel, and sent in batches along several lines or channels. The output of an S/P converter cannot go any faster than the input, but the circuit is useful when it is necessary to interface between a serial-data device and a parallel-data device.

Figure 2 illustrates a communications circuit in which a P/S converter is used at the source and an S/P converter is used at the destination. In this example, the data characters are 8-bit bytes; the illustration shows the transfer of one character.

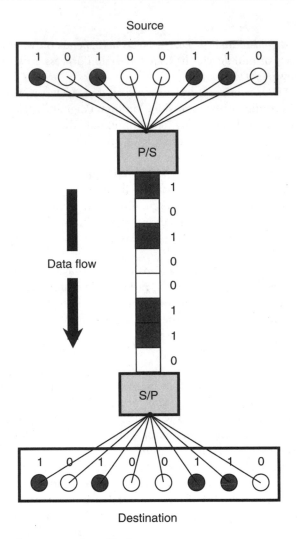

Data conversion—Fig. 2

DECIMAL NUMBER SYSTEM

See NUMERATION.

DEGREES OF FREEDOM

The term *degrees of freedom* refers to the number of different ways in which a robot arm can move. Most robot arms move in three dimensions, but they often have more than three degrees of freedom.

You can use your own arm to get an idea of the degrees of freedom that a robot arm might have. Extend your right arm straight out toward the horizon. Extend your index finger so it is pointing. Keeping your arm straight, move it from the shoulder. You can move your arm three ways. Up-and-down movement is called *pitch*. Movement to the right and left is *yaw*. You can also rotate your whole arm as if you were using it as a screwdriver; this is *roll*. Your shoulder has three degrees of freedom: pitch, yaw, and roll.

Now move your arm from the elbow only. If you hold your shoulder and upper arm in the same position constantly, you can see that your elbow joint has the equivalent of pitch in your shoulder joint. But that is all (unless your elbow is dislocated). The human elbow has one degree of freedom.

Extend your arm toward the horizon, straighten it out, and move only your wrist. Keep the arm above the wrist straight and motionless. Your wrist can bend up and down and can also move side to side. The human hand has two degrees of freedom with respect to the arm above it: *pitch* and *yaw*. Thus, in total, your shoulder/elbow/wrist system has six degrees of freedom: three in the shoulder, one in the elbow, and two in the wrist. A certain amount of *roll* is also possible in the arm below the elbow; this does not occur in either the elbow joint or the wrist joint, but in the lower arm itself. This makes for a seventh degree of freedom.

Three degrees of freedom are sufficient to bring the end of a robot arm to any point within its *work envelope*, or work space, in three dimensions. Thus, in theory, it might seem that a robot should never need more than three degrees of freedom. But the extra possible motions, provided by multiple joints, give robot arms versatility that they could not have with only three degrees of freedom.

See also ARTICULATED GEOMETRY, CARTESIAN COORDINATE GEOMETRY, CYLINDRICAL COORDINATE GEOMETRY, DEGREES OF ROTATION, POLAR COORDINATE GEOMETRY, REVOLUTE GEOMETRY, ROBOT ARM, SPHERICAL COORDINATE GEOMETRY, and WORK ENVELOPE.

DEGREES OF ROTATION

Degrees of rotation are a measure of the extent to which a robot joint, or a set of robot joints, is turned. Some reference point is always used, and the angles are specified in degrees or radians with respect to that joint.

Rotation in one direction (usually clockwise) is represented by positive angles; rotation in the opposite direction is specified by negative angles. Thus, if angle $X = 58°$, it refers to a rotation of $58°$ clockwise with respect to the reference axis. If angle $Y = -74°$, it refers to a rotation of $74°$ counterclockwise.

The illustration shows a robot arm with three joints. The reference axes are J1, J2, and J3, for rotation angles X, Y, and Z. The individual angles add together.

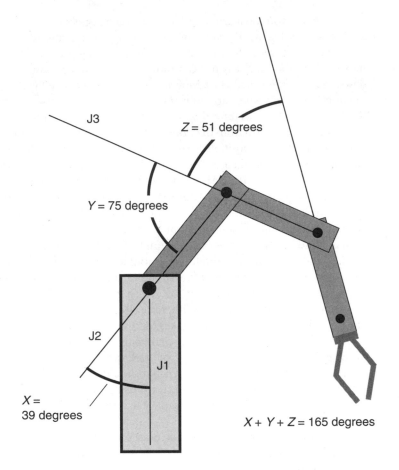

J3

$Z = 51$ degrees

$Y = 75$ degrees

J2

J1

$X = 39$ degrees

$X + Y + Z = 165$ degrees

Degrees of rotation

When it is necessary to move this robot arm to a certain position within its work envelope, or the region in space that the arm can reach, the operator enters data into a computer. This data includes the measures of angles X, Y, and Z. In the example shown by the illustration, the operator has specified $X = 39°$, $Y = 75°$, and $Z = 51°$. For simplicity, no other possible variable parameters, such as base rotation, wrist rotation, or extension/retraction of linear sections, are shown.

See also ARTICULATED GEOMETRY, DEGREES OF FREEDOM, ROBOT ARM, and WORK ENVELOPE.

DELIBERATION

Deliberation refers to any characteristic of robotic navigation that involves advance planning of some sort, rather than mere reaction to the presence of obstacles or changes in the work environment. Deliberative planning is commonly combined with another scheme called *reactive planning*.

See also HIERARCHICAL PARADIGM, HYBRID DELIBERATIVE/REACTIVE PARADIGM, and REACTIVE PARADIGM.

DEPTH MAP

A *depth map*, also called a *range image*, is a specialized form of *computer map*, rendered as a grayscale image of a robot's work environment. The brightness of each pixel (picture element) in the image is proportional to the *range*, or radial distance, to the nearest obstruction in a specific direction. In some depth maps, the brightest pixels correspond to short range; in others, the brightest pixels correspond to long range.

A typical range image looks something like a grayscale video image or its negative. However, upon examination, the difference between a conventional visible or infrared (IR) image and a depth map becomes apparent. Local detail in objects, such as the contour of a human face, generally do not show up in a depth map, even if the shade, color, or heat radiation vary greatly. It is the radial distance, as determined by a *range sensing and plotting* system, which produces the image.

Suppose a robot is navigating across a flat field or empty parking lot on which a huge ball sits. The range sensing and plotting system is programmed to produce a depth map. In the field of view of the system, the only objects that appear are the flat surface and the ball. Suppose the depth map is such that the relative brightness of the image is inversely proportional to the radial distance. The depth map looks like the rendition shown in the accompanying illustration. The color of the ball and the surface on

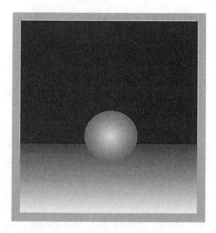

Depth map

67

which it rests, and the time of day or night, do not matter; the rendition is based entirely on the range as a function of the direction in three-dimensional space.

See also **COMPUTER MAP** and **RANGE SENSING AND PLOTTING.**

DERIVATIVE

The term *derivative* refers to the rate of change of a mathematical *function*. For example, speed or velocity is the derivative of displacement, and acceleration is the derivative of velocity.

Figure 1 shows a hypothetical graph of displacement as a function of time. This function appears as a curve. You might think of it as a graph of the distance traveled by a robot accelerating along a linear track, with the displacement specified in meters and the time in seconds. At any specific instant in time, call it *t*, the speed is equal to the slope of the line tangent to the curve at that moment. This quantity is expressed in linear displacement units (such as meters) per second.

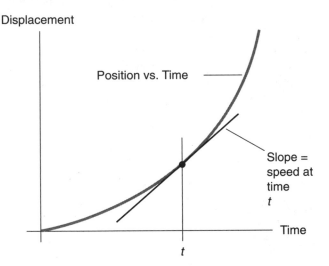

Derivative—Fig. 1

In digital electronics, a circuit that continuously takes the derivative of an input wave, as a function of instantaneous amplitude versus time, is called a *differentiator*. An example of the operation of a differentiator is shown in Fig. 2. The input is a *sine wave*. The output follows the slope, or derivative, of this wave; the result is a *cosine wave*, with the same shape as the sine wave but displaced by one-fourth of a cycle (90° of phase). Compare **INTEGRAL.**

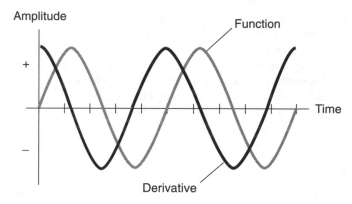

Derivative—Fig. 2

DICHOTOMIZING SEARCH

See BINARY SEARCH.

DIFFERENTIAL AMPLIFIER

A *differential amplifier* is an electronic circuit that responds to the difference in amplitude between two signals. Some differential amplifiers also produce gain, resulting in an output signal whose amplitude varies dramatically when the amplitude of either input signal varies only a little. The output is proportional to the difference between the input signal levels. If the input amplitudes are identical, then the output is zero.

The nomograph shows how the instantaneous output of a differential amplifier varies as the instantaneous input values change. To find the

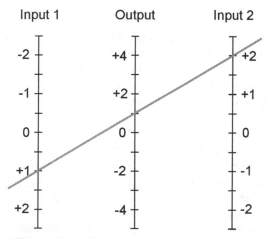

Differential amplifier

output, place a straight ruler so its edge passes through the two input points; the output is the point on the center scale through which the ruler passes. In this example, the circuit has no gain.

Differential amplifiers are sometimes employed in robotic sensing systems. The output of an amplifier in this situation can be used as an error signal, which is sent to the guidance system to regulate the movement of a mobile robot. This can ensure that the robot follows a prescribed route in its work environment, such as the path along which two reference acoustic or radio waves are exactly in phase. Compare DIFFERENTIAL TRANSDUCER.

DIFFERENTIAL TRANSDUCER

A *differential transducer* is a sensing device with two inputs and one output. The output is proportional to the difference between the input signal levels. An example is a *differential pressure transducer,* which responds to the difference in mechanical pressure at two points.

Any pair of transducers can be connected in a differential arrangement. Usually, this involves connecting the transducers to the inputs of a *differential amplifier.*

When the two variables have the same magnitude, the output of the differential transducer is zero. The greater the difference in the magnitudes of the sensed effects, the greater is the output. The most output occurs when one of the sensed effects is intense, and the other is zero or near zero. Whether the output is positive or negative depends on which of the sensed effects is greater. Compare DIFFERENTIAL AMPLIFIER.

DIFFERENTIATION

See DERIVATIVE.

DIGITAL IMAGE

A *digital image,* also called a *digitized image,* is a rendition of a scene at visible, infrared (IR), or ultraviolet (UV) wavelengths, or using *radar* or *sonar,* in the form of a rectangular array of tiny squares or dots called *pixels.*

In a *grayscale digital image,* each pixel has a brightness level that can attain any of numerous discrete binary values. Common ranges are from binary 0000 through 1111 (16 shades of gray) or binary 00000000 through 11111111 (256 shades of gray).

In a *color digital image,* each pixel has a color value of red, green, or blue (RGB), and also a brightness level that can attain any of numerous discrete binary values. Color digital images occupy considerably more data memory or storage space than grayscale digital images, because the three color values can vary independently with each pixel.

The number of pixels in a digital image determines the *resolution.* This figure is generally represented in terms of the number of pixels in the horizontal and vertical dimensions. In a computer display, for example, a common resolution is 1024×768 (1024 pixels horizontal and 768 pixels vertical).

In a visible digital image, color is usually rendered in as true-to-life a fashion as possible. However, at IR and UV wavelengths, and especially with radar and sonar, false colors are often used in digital images. For example, in a sonar image, color can represent the *range,* or the distance between a robot and objects in its work environment. Red might represent the smallest range, progressing up through orange, yellow, green, blue, violet, and finally white, representing the greatest (or infinite) range.

See also RESOLUTION.

DIGITAL INTEGRATED CIRCUIT

See INTEGRATED CIRCUIT.

DIGITAL LOGIC

See LOGIC GATE.

DIGITAL MOTION

In robotics, *digital motion* refers to the movement of a robot arm that can stop only at certain positions within its work envelope. This is in contrast to *analogical motion,* in which the number of possible positions is theoretically infinite.

The possible positions in a system that incorporates digital motion must be programmed into the robot controller. For example, the base of a robot arm might rotate to any multiple of 30° in the complete circle of 0° to 360°. This allows for 12 unique base-rotation angles. If more precision is required, the angle increment can be reduced (10° will allow for 36 unique base-rotation angles, for example). When the robot arm must be rotated to a certain base angle position, the desired angle or step is entered into the robot controller. The arm then moves to the designated position and stops.

Stepper motors are commonly used in robotic digital-motion systems. These motors move in discrete increments, rather than rotating continuously. Compare ANALOGICAL MOTION.

See also STEPPER MOTOR.

DIGITAL SIGNAL PROCESSING (DSP)

Digital signal processing (DSP) is a scheme for improving the precision of digital data. It can be used to clarify or enhance signals of all kinds.

Analog cleanup

When DSP is used in an analog communications system, the signal is first changed into digital form by *A/D conversion.* Then the digital signal is "tidied up" so the pulse timing and amplitude adhere strictly to protocol. Finally, the digital signal is changed back to analog form by means of *D/A conversion.*

Digital signal processing can extend the workable range of a communications circuit, because it allows reception under worse conditions than would be possible without it. Digital signal processing also improves the quality of fair signals, so the receiving equipment or operator makes fewer errors. The DSP process also ensures that the necessary communications bandwidth is kept to a minimum.

Digital cleanup

In circuits that use only digital modes, A/D and D/A conversion are irrelevant, but DSP can nevertheless "tidy up" the signal. This improves the accuracy of the system, and also makes it possible to copy data many times (that is, to produce multigeneration copies).

The DSP circuit minimizes confusion between digital states, as shown in the illustration. A hypothetical signal before processing is shown at the top; the signal after processing is shown at the bottom. If the input amplitude is above a certain level for an interval of time, the output is high (logic 1). If the input amplitude is below the critical point for a time interval, then the output is low (logic 0). A strong burst of noise might

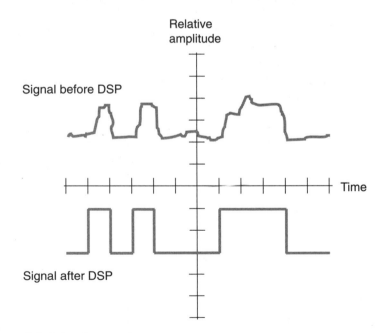

Digital signal processing

fool the circuit into thinking the signal is high when it is actually low; but overall, errors are less frequent with DSP than without it.

In computers and robots

A DSP system can be etched onto a single *integrated circuit (IC)*, similar in size to a memory chip. Some DSP circuits serve multiple functions in a computer or robotic system, so the controller can devote itself to doing its primary work without having to bother with extraneous tasks.

A DSP chip can compress and decompress data, help a computer recognize and generate speech, translate from one spoken language to another (such as from English to Chinese or vice versa), and recognize and compare patterns.

See also DATA CONVERSION.

DIRECTIONAL TRANSDUCER

A *directional transducer* is a device that senses some effect or disturbance, and produces an output signal that varies in amplitude depending on the direction from which the effect or disturbance arrives. Directional transducers are used extensively in robotic sensing and guidance systems.

A simple example of a directional transducer is a common microphone. Microphones are almost always *unidirectional,* that is, they respond best in one direction. An example of a *bidirectional transducer* is a horizontal radio antenna known as a *dipole.* Some transducers are *omnidirectional* in a specified plane. A vertical radio antenna is an example. It works equally well in all horizontal directions. However, its sensitivity varies in vertical planes. Some transducers are equally sensitive in all possible directions; the directional pattern for such a device is a sphere in three dimensions. This is a truly omnidirectional transducer.

DIRECTION FINDING

Direction finding is a means of location and/or navigation, usually employing radio or acoustic waves. At radio frequencies (RF), location and navigation systems operate between a few kilohertz and the microwave region. Acoustic systems use frequencies between a few hundred hertz and a few hundred kilohertz.

Signal comparison

A *mobile robot* can find its position by comparing the signals from two fixed stations whose positions are known, as shown in Fig. 1. By adding 180° to the bearings of the sources X and Y, the robot (square) obtains its bearings as "seen" from the sources (dots). The robot can determine its direction and speed by taking two readings separated by a certain amount of time. Computers can assist in precisely determining, and displaying, the position and the velocity vector.

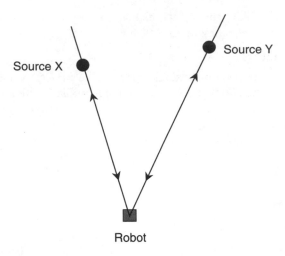

Direction finding—Fig. 1

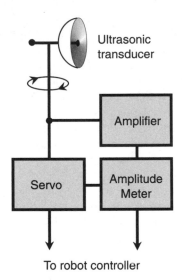

Direction finding—Fig. 2

Figure 2 is a block diagram of an *acoustic direction finder*. In this case the acoustic waves are ultrasound. The receiver has a signal-strength indicator and a *servo* that turns a directional ultrasonic transducer. There are two signal sources at different frequencies. When the transducer is rotated so the signal from one source is maximum, a bearing is obtained by comparing the orientation of the transducer with some known standard such as the reading of a magnetic compass. The same is done for the other source. A computer uses *triangulation* to figure out the precise location of the robot.

Radio direction finding (RDF)

A radio receiver, equipped with a signal-strength indicator and connected to a rotatable, directional antenna, can be used to determine the direction from which signals are coming. *Radio direction finding* (RDF) equipment aboard a mobile robot facilitates determining the location of a transmitter. RDF equipment can be used to find the location of a robot with respect to two or more transmitters operating on different frequencies.

In an RDF receiver, a *loop antenna* is generally used. It is shielded against the electric component of radio waves, so it picks up only the magnetic flux. The circumference is less than 0.1 wavelength. The loop is rotated until a dip occurs in the received signal strength. When the dip is found, the axis of the loop lies along a line toward the transmitter. When readings are taken from two or more locations separated by a sufficient distance, the transmitter can be pinpointed by finding the intersection point of the azimuth bearing lines on a map.

At frequencies above approximately 300 MHz, a directional transmitting/receiving antenna, such as a *Yagi, quad, dish,* or *helical* type, gives better results than a small loop. When such an antenna is employed for RDF, the azimuth bearing is indicated by a signal peak rather than by a dip.

See also DIRECTION RESOLUTION and TRIANGULATION.

DIRECTION RESOLUTION

Direction resolution refers to the ability of a robot to separate two objects that appear, from the robot's point of view, to lie in almost the same direction. Direction resolution on the Earth's surface is also called *azimuth resolution.* Quantitatively, it is specified in degrees, minutes, or seconds of arc.

Two objects might be so nearly in the same direction that a robot "sees" them as being one and the same object, but if they are at different radial distances, the robot can tell them apart by distance measurement.

See also DIRECTION FINDING, DISTANCE MEASUREMENT, RADAR, and SONAR.

DISPLACEMENT ERROR

Displacement error refers to an imprecision in robot position that takes place over time. Displacement error can be measured in absolute terms, such as linear units or degrees of arc. It can also be measured in terms of a percentage of the total displacement or rotation.

As an example, suppose a mobile robot is programmed to proceed at a speed of 1.500 meters per second (m/s) at an azimuth bearing of 90.00° (due east) on a level surface. After 10 s, this robot can be expected to be 15.00 m due east of its starting position. If the robot encounters an upward incline, the displacement might be less than 15.00 m; if the robot encounters a downslope, the displacement might be more. If the surface banks to the left or the right, the direction of motion can be expected to change, causing the robot to end up to the north or south of its position had it traveled on a level surface. In the ideal scenario, terrain irregularity would not affect the speed or direction of the machine; the displacement error would therefore be zero.

Displacement errors can result from the accumulation of *kinematic error* over time. Compare KINEMATIC ERROR.

DISPLACEMENT TRANSDUCER

A *displacement transducer* is a device that measures a distance or angle traversed, or the distance or angle separating two points. Some displacement transducers convert an electrical current or signal into movement over a certain distance or angle. A transducer that measures distance in a straight line is a *linear displacement transducer.* If it measures an angle, it is an *angular displacement transducer.*

Suppose you want a robot arm to rotate 28° in the horizontal plane—no more and no less. You give a command to the robot controller such as "BR = 28" (base rotation = 28°). The controller sends a signal to the robot arm, so that it rotates clockwise. An angular displacement transducer keeps track of the angle of rotation, sending a signal back to the computer. This signal increases in linear proportion to the angle that the arm has turned. By issuing the command "BR = 28," you tell the controller two things:

1. Start the base of the arm rotating.
2. Stop the rotation when the arm has turned through 28°.

The second component of the command sets a threshold level for the return signal. As the signal from the displacement transducer increases, it reaches this threshold at 28° of rotation. The controller is programmed to stop the arm at this time.

There are other ways to get a robot arm to move, besides using displacement transducers. The above is just one example of how such a transducer might be used in a robotic system.

See also ROBOT ARM and TEACH BOX.

DISTANCE MEASUREMENT

Distance measurement, also called *ranging,* is a scheme that an *autonomous robot* can use to navigate in its work environment. It also allows a central computer to keep track of the whereabouts of *insect robots*. There are several ways for an autonomous robot to measure the distance between itself and some object.

Sonar uses sound or ultrasound, bouncing the waves off of things around the robot and measuring the time for the waves to return. If the robot senses that an echo delay is extremely short, it knows that it is getting too close to something. Acoustic waves propagate at a speed of roughly 335 m/s in dry air at sea level.

Radar works like sonar, but uses microwave radio signals rather than sound waves. Light beams can also be used, particularly lasers, in which case the scheme is called *ladar*. But radio and light beams travel at such high speed (300 million m/s in free space) that it is difficult to measure delay times for nearby objects. Also, some objects reflect light waves poorly, making it difficult to obtain echoes strong enough to allow distance measurement.

Stadimetry infers the distance to an object of known height, width, or diameter by measuring the angle the object subtends in the vision system's field of view.

Beacons of various kinds can be used for distance measurement. These devices can use sound, radio waves or light waves.

See also AUTONOMOUS ROBOT, BEACON, DISTANCE RESOLUTION, INSECT ROBOT, LADAR, RADAR, SONAR, STADIMETRY, and TIME-OF-FLIGHT DISTANCE MEASUREMENT.

DISTANCE RESOLUTION

Distance resolution is the precision of a robotic *distance measurement* system. Qualitatively, it is the ability of the system to differentiate between two objects that are almost, but not quite, the same distance away from the robot. Quantitatively, it can be measured in meters, centimeters, millimeters, or even smaller units.

When two objects are very close to each other, a distance-measuring system sees them as a single object. As the objects get farther apart, they become distinguishable. The minimum radial separation of objects, for a ranging system to tell them apart, is the distance resolution.

With some distance measuring systems, nearby sets of objects can be resolved better than sets of objects far away. Suppose two objects are separated radially by 1 m. If their mean (average) distance is 10 m, their separation is 1/10 (10 percent) of the mean distance. If their mean distance is 1000 m, their separation is 1/1000 (0.1 percent) of the mean distance. If the distance resolution is 1 percent of the mean distance, then the system can tell the nearer pair of objects apart, but not the more distant pair.

Distance resolution depends on the type of ranging system used. The most sensitive methods compare the phases of the wavefronts emitted by laser beams. These waves either arrive from, or are reflected by, *beacons* located at strategic points in the work environment. A high-end system of this kind can resolve distances down to a small fraction of a millimeter.

See also BEACON, DISTANCE MEASUREMENT, RADAR, and SONAR.

DISTINCTIVE PLACE

A *distinctive place* is a point in a mobile robot's work environment that has special significance, or that can be used as a point of reference for navigational purposes. Such points are determined on the basis of features in specific regions, called *neighborhoods*, in the work environment.

Suppose a mobile robot is designed to function on a single level of an office building. The work environment is the entire floor (the set of all points) over which the machine can move. Each room can be considered a neighborhood. Distinctive places might be defined as the centers of the doorways between adjacent rooms, or between each room and the hallway. Distinctive places might also include the physical (geographic) center of the floor in each room, or the point on the floor that lies at the greatest distance, in any given room, from fixed obstructions. *Beacons* can also serve as distinctive places.

See also BEACON, COMPUTER MAP, and RELATIONAL GRAPH.

DISTRIBUTED CONTROL

In a system containing more than one robot, *distributed control* refers to unit independence. In a robotic system that employs distributed control, also known as decentralized control, each robot in the fleet is capable, to some extent, of making its own decisions and operating without instructions from other robots or from a central controller. If there is a central controller, its function is limited. This type of robotic system is analogous to a peer-to-peer computer network.

In a robotic system that employs *uniformly distributed control,* there is no main controller; each robot is fully autonomous, containing its own controller. Each unit is equal to all the others in significance. In some systems, there is a main controller that oversees some of the operations of each unit in the fleet. This is known as *partially distributed control.* Another example of partially distributed control is a system in which each robot receives a set of instructions from a central controller, stores those instructions, and then carries them out independently of the central controller.

In some robotic systems, the individual units are completely and continuously dependent on the central controller, and cannot function if the communication link is severed. Such a system is said to employ *fully centralized control.* Compare CENTRALIZED CONTROL.

See also AUTONOMOUS ROBOT and INSECT ROBOT.

DOMAIN OF FUNCTION

The *domain* of a mathematical *function* is the set of independent-variable values for which the function is defined. Every x in the domain of a function f is mapped by f onto a definite, single value y. Any x not in the domain is not mapped onto anything by the function f.

Suppose you are given the function $f(x) = +x^{1/2}$ (that is, the positive square root of x). The graph of this function is shown in the illustration. The function is not defined for negative values of x, and is also not defined, as shown in this particular example, for $x = 0$. The function $f(x)$ has

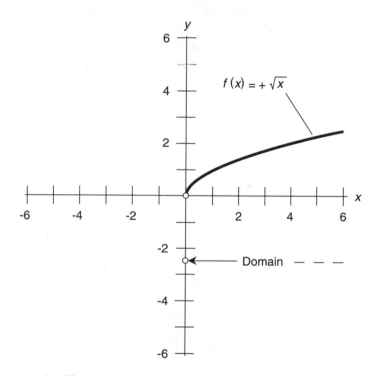

Domain of function

values y if, and only if, $x > 0$. Therefore the domain of f is the set of positive real numbers.

Computers work extensively with functions, both analog and digital. Functions are important in robotic navigation, location, and measurement systems.

See also **FUNCTION** and **RANGE OF FUNCTION**.

DOWNLINK

See UPLINK/DOWNLINK.

DROP DELIVERY

Drop delivery is a simple method that a robotic *end effector* can use to place an object into position. The object is picked up by a gripper and then moved until it is directly over a slot, hole, conveyor belt, chute, or other receptacle designed for it. Then the gripper lets go of the object, and it falls into place.

Drop delivery requires precision in the movement of the robotic arm and end effector. In addition, when the gripper lets go of the object, it must not impart significant lateral force or torque to the object. Otherwise the object might move out of alignment or topple. If a conveyor belt is used, some means must be employed to ensure that the movement of the belt does not cause the object to slip, tip over, or fall off the belt after it lands.

DROPOFF

See MAGNITUDE PROFILE.

DUTY CYCLE

The *duty cycle* is the proportion of time during which a circuit, machine, or component is operated.

Suppose a motor is run for 1 min, then is shut off for 2 min, then is run for 1 min again, and so on. The motor therefore runs for 1 out of every 3 min, or one-third of the time. Its duty cycle is therefore ⅓, or 33 percent.

If a device is observed for a length of time t_o, and during this time it runs for a total time t (in the same units as t_o), then the duty cycle expressed as a percentage, $d_\%$, is given by the following formula:

$$d_\% = \frac{100t}{t_o}$$

When determining the duty cycle, it is important that the observation time be long enough. In the case of the motor described above, any value of less than 3 min is too short to get a complete sample of the data. Ideally, the observation time should be at least twice the time required for a complete cycle of activity. If the cycle of activity varies somewhat (a common situation), then the observation time must be much greater than the time required for a single cycle.

The more a circuit, machine, or component is used, the sooner it will wear out, if all other factors are held constant. In general, the higher the duty cycle, the shorter is the useful life. This effect is most pronounced when a device is worked near its limits. Also, the rating of a device often depends on the duty cycle at which it is expected to be used.

Suppose the motor described above is rated at a torque of 10 newton-meters (10 N·m) for a duty cycle of 100 percent. If the motor is called upon to provide a constant torque of 9.9 N·m, then it will be taxed to its utmost. If it must constantly turn a load of 12 N·m, it should come as no surprise if it fails prematurely. For a duty cycle of 33 percent, the motor might be rated at 15 N·m, as long as any single working period does not exceed 2 min. If it only needs to turn 0.5 N·m, the motor can not only run continuously, it will probably last longer than its expected life.

Devices such as robot motors can be protected from overwork (either momentary or long-term) by means of *back pressure sensors*. See BACK PRESSURE SENSOR.

DYNAMIC STABILITY

Dynamic stability is a measure of the ability of a robot to maintain its balance while in motion.

A robot with two or three legs, or that rolls on two wheels, can have excellent stability while it is moving, but when it comes to rest, it is unstable. A two-legged robot can be pushed over easily when it is standing still. This is one of the major drawbacks of biped robots. It is difficult and costly to engineer a good sense of balance, of the sort you take for granted, into a two-legged or two-wheeled machine, although it has been done.

Robots with four or six legs have good dynamic stability, but they are usually slower in their movements compared with machines that have fewer legs.

See also BIPED ROBOT, INSECT ROBOT, and STATIC STABILITY.

DYNAMIC TRANSDUCER

A *dynamic transducer* is a coil-and-magnet device that converts mechanical motion into electricity or vice versa. The most common examples are the *dynamic microphone* and the *dynamic loudspeaker*. Dynamic transducers can be used as sensors in a variety of robotic applications.

The illustration is a functional diagram of a dynamic transducer suitable for converting sound waves into electric currents or vice versa. A diaphragm is attached to a permanent magnet. The magnet is surrounded by a coil of wire. Acoustic vibrations cause the diaphragm to move back and forth; this moves the magnet, which causes fluctuations in the magnetic field within the coil. The result is alternating-current (AC) output from the coil, having the same waveform as the sound waves that strike the diaphragm.

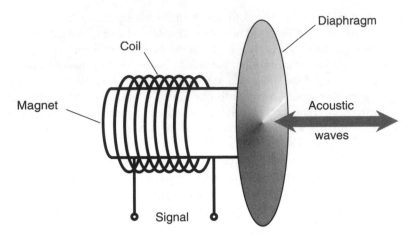

Dynamic transducer

If an audio signal is applied to the coil of wire, it creates a magnetic field that produces forces on the permanent magnet. This causes the magnet to move, pushing the diaphragm back and forth. This displaces the air near the diaphragm, producing acoustic waves that follow the waveform of the signal.

Dynamic transducers are commonly used in robotic *speech recognition* and *speech synthesis* systems. Compare ELECTROSTATIC TRANSDUCER and PIEZOELECTRIC TRANSDUCER.

See also SPEECH RECOGNITION and SPEECH SYNTHESIS.

E

EDGE DETECTION

Edge detection is to the ability of a robotic *vision system* to locate boundaries. It also refers to a robot's knowledge of what to do with respect to those boundaries.

A robot car, for example, uses edge detection to see the edges of a road, and uses the data to keep itself on the road. However, it also needs to stay a certain distance from the right-hand edge of the pavement, so that it does not cross over into the lane of oncoming traffic. It must stay off the road

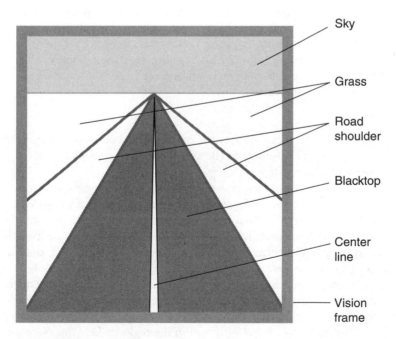

Edge detection

shoulder. Thus, it must tell the difference between pavement and other surfaces, such as gravel, grass, sand, and snow. The robot car can use beacons for this purpose, but this requires the installation of the guidance system beforehand, limiting the robot car to roads that are equipped with such navigation aids.

A personal robot equipped with edge detection can see certain contours in its work environment. This keeps the machine from running into walls, or closed doors, or windows, or from falling down stairwells. Compare EMBEDDED PATH.

See also VISION SYSTEM.

EDUCATIONAL ROBOT

The term *educational robot* applies to any robot that causes its user(s) to learn something. Especially, this term applies to robots available for consumer use. Robots of this kind have become popular among children, particularly in Japan, but increasingly in the United States and other Western nations. These machines are toys in the sense that children have fun using them, but they often are excellent teachers as well. Children learn best when they are having fun at the same time.

An instructional robot is an educational robot intended to function only, or primarily, as a teacher. Robots of this kind can be purchased for use in the home, but more often they are found in schools, especially at the junior-high and senior-high levels (grades 7 through 12).

Robots are intimidating to some students. But once a child or young adult gets used to working or playing with machines, robots can become companions, especially if there is some measure of *artificial intelligence (AI)*.

See also PERSONAL ROBOT.

ELASTOMER

An *elastomer* is a flexible substance resembling rubber or plastic. In robotic *tactile sensing,* elastomers can be used to detect the presence or absence of mechanical pressure.

The illustration shows how an elastomer can be used to detect, and locate, a pressure point. The elastomer conducts electricity fairly well, but not perfectly. It has a foamlike consistency, so it can be compressed. An array of electrodes is connected to the top of the elastomer pad; an identical array is connected to the bottom of the pad. These electrodes run to the robot controller.

When pressure appears at some point in the elastomer pad, the material compresses, and this lowers its electrical resistance in a small region. This is detected as an increase in the current between electrodes in the top pad and the bottom pad, but only within the region where the elastomer is

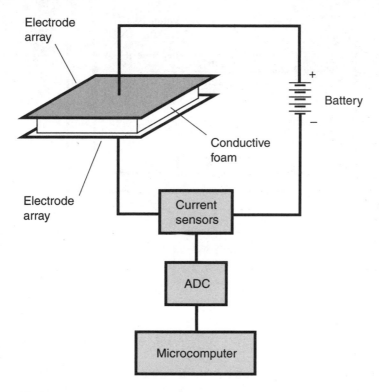

Elastomer

being compressed. The data is sent to an *analog-to-digital converter (ADC)* and then to a microcomputer, which determines where the pressure is taking place, and how intense it is.

See also TACTILE SENSING.

ELECTRIC EYE

An *electric eye* optically senses an object and then actuates a device. For example, it might be set up to detect anything passing through a doorway. This can count the number of people entering or leaving a building. Another example is the counting of items on a fast-moving assembly line; each item breaks the light beam once, and a circuit counts the number of interruptions.

Usually, an electric eye has a light source and a photocell; these are connected to an actuating circuit as shown in the block diagram. When something interrupts the light beam, the voltage or current from the photocell changes dramatically. It is easy for electronic circuits to detect

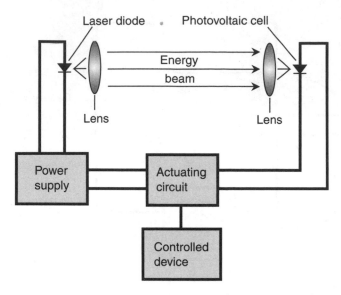

Electric eye

this voltage or current change. Using amplifiers, even the smallest change can be used to control large machines.

Electric eyes do not always operate with visible light. Infrared (IR), with a wavelength somewhat longer than visible red, is commonly used in optical sensing devices. This is ideal for use in burglar alarms, because an intruder cannot see the beam, and therefore cannot avoid it.

ELECTROCHEMICAL POWER

An *electrochemical cell* is a unit source of direct-current (DC) power. When two or more such cells are connected in series to increase the voltage, the result is a *battery*. Electrochemical cells and batteries are extensively used in mobile robots.

Lead–acid cell

Figure 1 shows an example of a *lead–acid cell*. An electrode of lead and an electrode of lead dioxide, immersed in a sulfuric acid solution, exhibit a potential difference. This voltage can drive a current through a load. The maximum available current depends on the volume and mass of the cell.

If this cell is connected to a load for a long time, the current will gradually decrease, and the electrodes will become coated. The nature of the acid will change. All the potential energy in the acid will be converted into DC electrical energy, and ultimately into heat, visible light, radio waves, sound, or mechanical motion.

Primary and secondary cells

Some cells, once their chemical energy has all been changed to electricity and used up, must be discarded. These are *primary cells*. Other kinds of cells, like the lead–acid unit described above, can get their chemical energy back again by means of recharging. Such a component is a *secondary cell*.

Primary cells contain a dry electrolyte paste along with metal electrodes. They go by names such as *dry cell, zinc–carbon cell,* or *alkaline cell.* These cells are commonly found in supermarkets and other stores. Some secondary cells can also be found at the consumer level. *Nickel–cadmium (Ni–Cd or NICAD) cells* are one common type. These cost more than ordinary dry cells, and a charging unit also costs a few dollars. However, these rechargeable cells can be used hundreds of times, and can pay for themselves and the charger several times over.

An *automotive battery* is made from secondary cells connected in series. These cells recharge from an alternator or from an outside charging unit. This type of battery has cells like the one shown in Fig. 1. It is dangerous to short-circuit the terminals of such a battery because the acid can boil out. In fact, it is unwise to short-circuit any cell or battery, because it might explode or cause a fire.

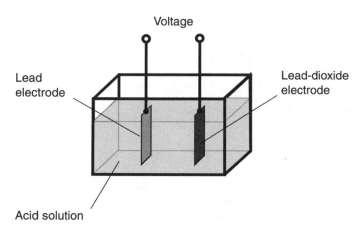

Electrochemical power—Fig. 1

Storage capacity

Common units of electrical energy are the *watt hour* (Wh) and the *kilowatt hour* (kWh). Any cell or battery has a certain amount of electrical energy that can be specified in watt hours or kilowatt hours. Often it is given in terms of the mathematical integral of deliverable current with respect to time, in units of *ampere hours* (Ah). The energy capacity in watt hours is the ampere-hour capacity multiplied by the battery voltage.

A battery with a rating of 20 Ah can provide 20 A for 1 h, or 1 A for 20 h, or 100 mA (100 milliamperes) for 200 h. The limitations are *shelf life* at one extreme, and *maximum deliverable current* at the other. Shelf life is the length of time the battery will remain usable if it is never connected to a load; this is measured in months or years. The maximum deliverable current is the highest current a battery can drive through a load without the voltage dropping significantly because of the battery's own internal resistance.

Small cells have storage capacity of a few milliampere hours (mAh) up to 100 or 200 mAh. Medium-sized cells might supply 500 mAh to 1000 mAh (1 Ah). Large automotive lead–acid batteries can provide upwards of 100 Ah.

Discharge curve

When an *ideal cell* or *ideal battery* is used, it delivers a constant current for a while, and then the current starts to decrease. Some types of cells and batteries approach this ideal behavior, exhibiting a *flat discharge curve* (Fig. 2). Others have current that decreases gradually from the beginning of use; this is a *declining discharge curve* (Fig. 3).

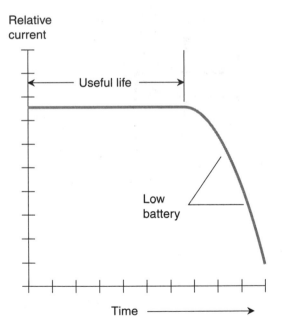

Electrochemical power—Fig. 2

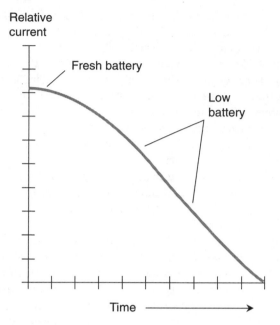

Relative current

Fresh battery

Low battery

Time ⟶

Electrochemical power—Fig. 3

When the current that a battery can provide has decreased to about half of its initial value, the cell or battery is said to be "weak" or "low." At this time, it should be replaced. A battery should not be allowed to run down until the current drops to nearly zero.

Common cells and batteries

The cells sold in stores, and used in convenience items such as flashlights and transistor radios, are usually of the zinc–carbon or alkaline variety. These provide 1.5 volts (V) and are available in sizes AAA (very small), AA (small), C (medium), and D (large). Batteries made from these cells are usually rated at 6 V or 9 V.

Zinc–carbon cells have a fairly long shelf life. The zinc forms the outer case and is the negative electrode. A carbon rod serves as the positive electrode. The electrolyte is a paste of manganese dioxide and carbon. Zinc–carbon cells are inexpensive and are usable at moderate temperatures, and in applications where the current drain is moderate to high. They do not work well in extremely cold environments.

Alkaline cells have granular zinc for the negative electrode, potassium hydroxide as the electrolyte, and a polarizer as the positive electrode. An alkaline cell can work at lower temperatures than a zinc–carbon cell. It also lasts longer in most electronic devices, and is therefore preferred for use in transistor radios, calculators, and portable cassette players. Its shelf life is much longer than that of a zinc–carbon cell.

Transistor batteries are small, 9-V, box-shaped batteries with clip-on connectors on top. They consist of six tiny zinc–carbon or alkaline cells in series. These batteries are used in low-current electronic devices, such as portable earphone radios, radio garage-door openers, television and stereo remote-control boxes, and electronic calculators.

Lantern batteries are rather massive, and can deliver a fair amount of current. One type has spring contacts on the top. The other type has thumbscrew terminals. Besides keeping an incandescent bulb lit for a while, these batteries, usually rated at 6 V and consisting of four zinc–carbon or alkaline cells, can provide enough energy to operate a low-power communications radio or a small mobile robot.

Silver-oxide cells are usually made into a buttonlike shape, and can fit inside a wristwatch. They come in various sizes and thicknesses, all with similar appearance. They supply 1.5 V and offer excellent energy storage for the weight. They have a flat discharge curve. Silver-oxide cells can be stacked to make batteries about the size of an AA cylindrical cell.

Mercury cells, also called *mercuric oxide cells,* have advantages similar to silver-oxide cells. They are manufactured in the same general form. The main difference, often not of significance, is a somewhat lower voltage per cell: 1.35 V. There has been a decrease in the popularity of mercury cells and batteries in recent years, because mercury is toxic and is not easily disposed of.

Lithium cells supply 1.5 to 3.5 V, depending on the chemistry used. These cells, like their silver-oxide cousins, can be stacked to make batteries. Lithium cells and batteries have superior shelf life, and they can last for years in very-low-current applications. They provide excellent energy capacity per unit volume.

Lead–acid cells and batteries have a solution or paste of sulfuric acid, along with a lead electrode (negative) and a lead-dioxide electrode (positive). Paste-type lead–acid batteries can be used in consumer devices that require moderate current, such as laptop computers, portable VCRs, and personal robots. They are also used in uninterruptible power supplies for personal computers.

Nickel-based cells and batteries

NICAD cells come in several forms. *Cylindrical cells* look like dry cells. *Button cells* are used in cameras, watches, memory backup applications, and other places where miniaturization is important. *Flooded cells* are used in heavy-duty applications, and can have a storage capacity of as much as 1000 Ah. *Spacecraft cells* are made in packages that can withstand extraterrestrial temperatures and pressures.

NICAD batteries are available in packs of cells that can be plugged into equipment to form part of the case for a device. An example is the battery pack for a hand-held radio transceiver.

NICAD cells and batteries should never be left connected to a load after the current drops to zero. This can cause the polarity of a cell, or of one or more cells in a battery, to reverse. Once this happens, the cell or battery will no longer be usable. When a NICAD nears full discharge, it should be recharged as soon as possible.

Nickel–metal-hydride (NiMH) cells and batteries can directly replace NICAD units in most applications.

See also POWER SUPPLY and SOLAR POWER.

ELECTROMAGNETIC SHIELDING

Electromagnetic shielding is a means of preventing computers and other sensitive equipment from being affected by stray *electromagnetic (EM) fields*. Computers also generate EM energy of their own, and this can cause interference to other devices, especially radio receivers, unless shielding is used.

The simplest way to provide EM shielding for a circuit is to surround it with metal, usually copper or aluminum, and to connect this metal to electrical ground. Because metals are good conductors, an EM field sets up electric currents in them. These currents oppose the EM field, and if the metal enclosure is grounded, the EM field is in effect shorted out. Interconnecting cables should also be shielded for optimum protection against *electromagnetic interference* (EMI). This is done by surrounding all the cable conductors with a copper braid. The braid is electrically grounded through the connectors at the ends of the cable.

One of the biggest advantages of fiber-optic data transmission is the fact that it does not need EM shielding. Fiber-optic systems are immune to the EM fields produced by radio transmitters and alternating-current (AC) utility wiring. Fiber-optic systems also work without generating external EM fields, so they do not cause EMI to surrounding circuits and devices.

ELECTROMECHANICAL TRANSDUCER

An *electromechanical transducer* is a device that converts electrical energy into mechanical energy or vice versa. *Electric motors* and *electric generators* are the most common examples. A motor works by means of magnetic forces produced by electric currents; the generator produces electric currents as a result of the motion of an electric conductor in a magnetic field.

Devices that convert sound into electricity, or vice versa, are another form of electromechanical transducer. *Speakers* and *microphones* are universal examples. They usually work by means of dynamic principles, but some work by electrostatic interactions.

Galvanometer-type analog meters, also known as *D'Arsonval meters*, are electromechanical transducers. They convert electric current into displacement. In recent years, digital meters have largely replaced electromechanical meters. Digital devices do not have moving parts to wear out, so they last much longer than electromechanical types. Digital meters are also able to tolerate more physical abuse.

Robots use electromechanical transducers in many ways. Examples include the *selsyn,* the *stepper motor,* and the *servomechanism.*

See also SELSYN, STEPPER MOTOR, and SERVOMECHANISM.

ELECTROSTATIC TRANSDUCER

An *electrostatic transducer* is a device that changes mechanical energy into electrical energy or vice versa by taking advantage of electrostatic forces. The most common types involve conversion between sound waves and audio-frequency electric currents.

The illustration is a functional diagram of an electrostatic transducer. It can function either as a microphone (sound-to-current transducer) or a speaker (current-to-sound transducer).

In the "microphone mode," incoming sound waves cause vibration of the flexible plate. This produces rapid (although small) changes in the

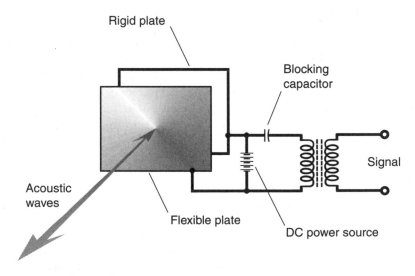

Electrostatic transducer

spacing, and therefore the capacitance, between the two plates. A direct-current (DC) voltage is applied to the plates, as shown. As the capacitance changes between the plates, the electric field between them fluctuates. This produces variations in the current through the primary winding of the transformer. Audio signals appear across the secondary winding.

In "speaker mode," currents in the transformer produce changes in the voltage between the plates. This change results in electrostatic force fluctuations, pulling and pushing the flexible plate in and out. The motion of the flexible plate produces sound waves.

Electrostatic transducers can be used in most applications where other types of transducers are employed. This includes speech recognition and speech synthesis systems. Advantages of electrostatic transducers include light weight and excellent sensitivity. They can also work with small electric currents. Compare DYNAMIC TRANSDUCER and PIEZOELECTRIC TRANSDUCER.

See also SPEECH RECOGNITION and SPEECH SYNTHESIS.

EMBEDDED PATH

An *embedded path* is a means of guiding a robot along a specific route. The *automated guided vehicle (AGV)* employs this scheme.

One common type of embedded path is a buried, current-carrying wire. The current in the wire produces a magnetic field that the robot can follow. This method of guidance has been suggested as a way to keep a car on a highway, even if the driver does not pay attention. The wire needs a constant supply of electricity for this guidance method to work. If the current is interrupted for any reason, the robot will lose its way.

Alternatives to wires, such as colored paints or tapes, do not need a supply of power, and this gives them an advantage. Tape is easy to remove and put somewhere else; this is difficult to do with paint, and practically impossible with wires embedded in concrete. Compare EDGE DETECTION.

See also AUTOMATED GUIDED VEHICLE.

EMPIRICAL DESIGN

Empirical design is an engineering technique in which experience and intuition are used in addition to theory. The process is largely trial and error. The engineer starts at a logical point, based on theoretical principles, but experimentation is necessary in order to get the device or system to work just right.

Robots are ideally suited to empirical design techniques. An engineer cannot draw up plans for a robot, no matter how detailed or painstaking the drawing-board process might be, and expect the real machine to work perfectly on the first trial. A prototype is built and tested, noting the flaws. The engineer goes back to the drawing board and revises the design. Sometimes it is necessary to start all over from scratch; more often, small changes are made. The machine is tested again, and the problems noted. Another drawing-board round follows. This process is repeated until the machine works the way the engineer (or the customer) wants.

END EFFECTOR

An *end effector* is a device or tool connected to the end of a *robot arm*. The nature of the end effects depends on the intended task.

If a robot is designed to set the table for supper, "hands," more often called *robot grippers,* can be attached to the ends of the robot arms. In an assembly-line robot designed to insert screws into cabinets, a rotating-shaft device and screwdriver head can be attached at the end of the arm. Such a rotating shaft might also be fitted with a bit for drilling holes, or an emery disk for sanding wood.

A given type of robot arm can usually accommodate only certain kinds of end effectors. One cannot take a table-setting robot, simply replace one of its grippers with a screwdriver, and then expect it to tighten the screws on the hinges of kitchen cabinets. Such a task change requires a change in the programming of the robot controller, so it operates in "handyrobot mode" rather than in "waitrobot mode." One must also change the hardware in the robot arm to operate a rotating end effector rather than a gripper.

See also ROBOT ARM and ROBOT GRIPPER.

ENTITIZATION

Entitization, also called *objectization,* is an expression of the ease with which a robot can differentiate among objects in its work environment. It is an indication of the effectiveness of *object recognition,* and can be defined in qualitative or quantitative terms.

Qualitative expressions of entitization are adjectives (such as "good," "fair," or "poor"). Quantitative entitization is determined on the basis of the proportion of correct identifications in a large number of tests in a practical scenario. For example, if a robot correctly identifies an object 997 out of 1000 times, its entitization is 99.7 percent accurate.

See also OBJECT RECOGNITION.

ENVELOPE

See WORK ENVELOPE.

EPIPOLAR NAVIGATION

Epipolar navigation is a means by which a machine can locate objects in three-dimensional (3-D) space. It can also navigate, and figure out its own position and path. Epipolar navigation works by evaluating the way an image appears to change as viewed from a moving point of view. The human eye/brain system does this to a limited degree with little thought or conscious effort. Robot *vision systems* can do it with extreme precision.

To illustrate epipolar navigation, imagine a robotic aircraft (drone) flying over the ocean. The only land beneath the drone is a small island (see the illustration). The robot controller has, on its hard drive, an excellent map that shows the location, size, and exact shape of this island. For instrumentation, the drone has only a computer, a good video camera, and sophisticated programming. The drone can navigate its way by

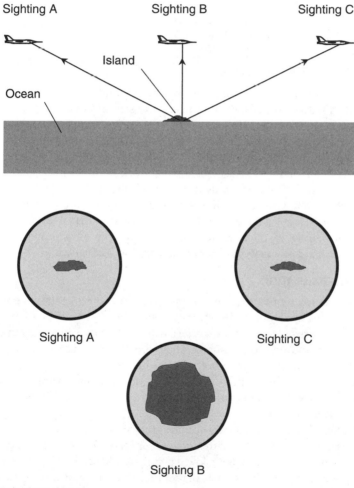

Epipolar navigation

observing the island and scrutinizing the shape and angular size of the island's image.

As the drone flies along, the island seems to move underneath it. A camera is fixed on the island. The controller sees an image that constantly changes shape and angular size. The controller is programmed with the true size, shape, orientation, and geographic location of the island. The controller compares the shape/size of the image it sees, from the vantage point of the aircraft, with the actual shape/size of the island, that it "knows" from the map data. From this alone, it can precisely determine the drone's:

- Altitude
- Speed of travel relative to the surface
- Direction of travel relative to the surface
- Geographic latitude
- Geographic longitude

The key is that there exists a one-to-one correspondence between all points within sight of the island, and the angular size and shape of the island's image. The correspondence is far too complex for a human being to memorize exactly; but for a computer, matching the image it sees with a particular point in space is easy.

Epipolar navigation can, in theory, work on any scale, and at any speed—even relativistic speeds at which time dilation occurs. It is a method by which robots can find their way without triangulation, direction finding, beacons, sonar, or radar. It is, however, necessary that the robot have a detailed, precise, and accurate computer map of its environment.

See also COMPUTER MAP, LOG POLAR NAVIGATION, and VISION SYSTEM.

ERROR ACCUMULATION

When measurements are made in succession, the maximum possible error adds up. This is called *error accumulation.*

Analog error accumulation can be illustrated by a measurement example. Suppose you want to measure a long piece of string (about 100 m, say), using a meter stick marked off in millimeters. You must place the stick along the string over and over again, about 100 times. If your error is up to ±2 mm with each measurement, then after 100 repetitions, the possible error is up to ±200 mm.

Digital error accumulation occurs as bits are misread in a communications circuit, incorrectly written on disk, or incorrectly stored in memory. A machine might see a logic low when it should see high, or vice versa. Suppose that, for a particular digital file, an average of three errors are introduced each time the file is transferred from one node to another in a communication circuit. If the signal passes through n nodes, there will be an average of $3n$ $(3 + 3 + 3 + \cdots + 3, n$ times) errors.

In robotic systems, *kinematic errors,* or errors in movement, can accumulate over time, resulting in eventual positioning or displacement errors.

See also KINEMATIC ERROR.

ERROR CORRECTION

Error correction is a form of computer programming in which certain types of mistakes are corrected automatically. An example is a program that maintains a large dictionary of English words. The operator of a computer connected to a speech-synthesizing robot might misspell words or make typographical errors. Running the error-correction program will cause the computer to single out all peculiar-looking words, bringing them to the attention of the operator. The operator can then decide whether the word is correct. With modern computers, huge vocabularies are easily stored.

When robots must keep track of variables such as position and speed, error correction can be used when an instrument is known to be imprecise, or when values depart from the reasonable range. A computer can keep track of *error accumulation,* checking periodically to be sure that discrepancies are not adding up beyond a certain maximum.

Error correction is important in robotic systems subject to *gravity loading.* In order to ensure that the *end effector* in a robot arm does not stray from its intended position because of the force of gravity on the assembly itself, *position sensing* devices can be used, and a feedback system employed to counter-move the robot arm until the error signal from the sensor is zero.

In robotic navigation systems, error correction refers to the set of processes that keep the device on its intended course. In a *servomechanism,* error correction is done by means of feedback.

See also ERROR ACCUMULATION, ERROR SENSING CIRCUIT, ERROR SIGNAL, POSITION SENSING, and SERVOMECHANISM.

ERROR-SENSING CIRCUIT

An *error-sensing circuit* produces a signal when two inputs are different, or when a variable deviates from a chosen value. If the two inputs are the same, or if the variable is at the chosen value, the output is zero. This type of circuit is also sometimes called a *comparator.*

Suppose you want a robot to home in on some object. The object has a radio transmitter that sends out a beacon signal. The robot has *radio direction finding* (RDF) equipment built-in. When the robot is heading in the right direction, the beacon is in the RDF null, and the received signal strength is zero, as shown in the accompanying polar-coordinate plot. If

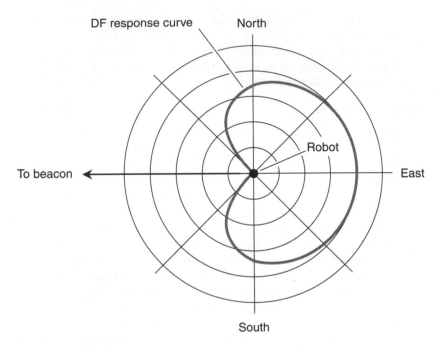

Error-sensing circuit

the robot turns off course, the beacon is no longer in the null, and a signal is picked up by the RDF receiver. This signal goes to the robot controller, which steers the robot to the left and right until the beacon signal once again falls into the null.

See also DIRECTION FINDING and SERVOMECHANISM.

ERROR SIGNAL

An *error signal* is a voltage generated by an *error-sensing circuit*. This signal occurs whenever the output of the device differs from a reference value. Error signals can be used in purely electronic systems, and also in electro-mechanical systems.

In the RDF device described under ERROR-SENSING CIRCUIT, the output might look like the polar-coordinate graph shown in the illustration. If the robot is pointed on course, the error signal is zero. If it is off course, either to the left or the right, a positive error signal voltage is generated, as shown in the accompanying rectangular-coordinate graph. The voltage depends on how far off course the robot is headed. In general, as the heading error increases, so does the error-signal strength.

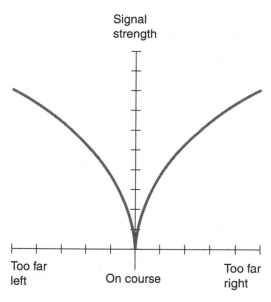

Signal
strength

Too far
left

On course

Too far
right

Error signal

A *direction-finding* circuit is designed to seek out, and maintain, a heading such that the error signal is always zero. To do this, the error signal is used by the robot controller to change the heading. This is the same principle by which a hidden radio transmitter is found.

See also BEACON, DIRECTION FINDING, ERROR CORRECTION, and SERVOMECHANISM.

EXOSKELETON

An *exoskeleton* is a robot arm that uses *articulated geometry* to mimic the motions of a human arm, and whose motions are controlled directly by movements of the arm of a human operator. Such devices can be used when working with hazardous materials. They are also useful as *prostheses* (artificial limbs). See ARTICULATED GEOMETRY and PROSTHESIS.

The term *exoskeleton* also refers to a specialized robot that is like a suit of armor a human can wear, and which can amplify movement displacement and/or force, resulting in physical strength far beyond that of an ordinary man or woman. A woman might, for example, lift a car over her head; the steel frame of the exoskeleton would bear the weight and pressure. A man might throw a baseball a kilometer. The armor could protect against blows, fire, and perhaps even bullets. *Full exoskeletons* have, to date, been implemented mainly in science-fiction stories.

A full exoskeleton differs from a *telepresence* system. In telepresence, the human operator is not at the same location as the robot. But when a human wears an exoskeleton, he or she is on site with the machine. This is both an asset and a liability: it allows for greater control and better sensing of the work environment, but in some instances it can place the human operator in physical peril. Compare TELEPRESENCE.

EXPERT SYSTEM

An *expert system* is a scheme of computer reasoning, also known as a *rule-based system*. Expert systems are used in the control of smart robots. They can also be employed in stand-alone computers.

The drawing is a block diagram of a typical expert system. The heart of the device is a set of facts and rules. In the case of a robotic system, the facts consist of data about the robot's environment, such as a factory, an office, or a kitchen. The rules are statements of the logical form "If X, then Y," similar to statements in high-level programming languages.

An *inference engine* decides which logical rules should be applied in various situations. Then it instructs the robot(s) to carry out certain tasks. However, the operation of the system can only be as sophisticated as the data supplied by human programmers.

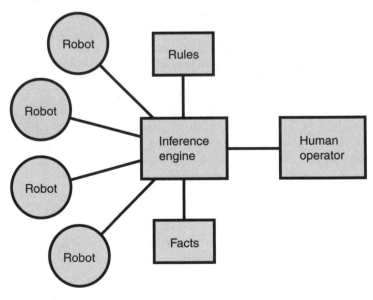

Expert system

Expert systems can be used in computers to help people do research, make decisions, and generate forecasts. A good example is a program that assists a physician in making a diagnosis. The computer asks questions, and arrives at a conclusion based on the answers given by the patient and doctor.

One of the biggest advantages of expert systems is the fact that reprogramming is easy. As the environment changes, the robot can be taught new rules, and supplied with new facts.

EXTENSIBILITY

Extensibility, also called *expandability,* refers to the ease with which a robotic system can be modified to perform a greater number, or a greater variety, of tasks than those allowed for in its original design.

The extensibility of a robotic system depends on various factors, including the nature of the hardware, the controller memory, the controller data storage space, and the controller processing speed. Extensibility is enhanced by the use of modular construction and standardized parts.

EXTRAPOLATION

When data are available within a certain range, an estimate of values outside that range can be made by a technique called *extrapolation.* This can be educated guessing, but it can also be done using a computer. The more sophisticated the computer software, the more accurately it can extrapolate.

An example of extrapolation is the forecast path of a hurricane as it approaches a coastline. Knowing its path up to the present moment, a range of possible future paths is developed by the computer. Factors that can be programmed into the computer to help it make an accurate extrapolation include:

- Paths of hurricanes in past years that approached in a similar way
- Steering currents in the upper atmosphere
- Weather conditions in the general path of the storm

The farther out (into the future) an extrapolation is made, the less accurate the results. While a weather computer might do a good job of predicting a hurricane path 24 h in advance, no machine yet devised can tell exactly where the storm will be in a week. Compare INTERPOLATION.

EYE-IN-HAND SYSTEM

For a *robot gripper* to find its way, a camera can be placed in the gripper mechanism. The camera must be equipped for work at close range, from about 1 m down to a few millimeters. The positioning error must be as small as possible, preferably less than 0.5 mm. To be sure that the camera gets a good image, a lamp is included in the gripper along with the camera (see the drawing).

The so-called *eye-in-hand system* can be used to measure precisely how close the gripper is to whatever object it is seeking. It can also make positive identification of the object, so that the gripper does not go after the wrong thing.

The eye-in-hand system uses a *servomechanism*. The robot is equipped with, or has access to, a controller that processes the data from the camera and sends instructions back to the gripper.

See also FINE MOTION PLANNING and ROBOT GRIPPER.

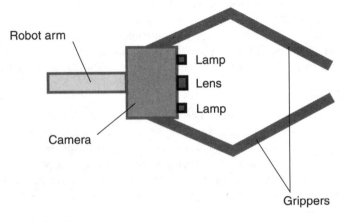

Eye-in-hand system

F

FALSE NEGATIVE OR POSITIVE

Sensors do not always react as intended to stimuli, or *percepts,* in the environment. This can occur for a variety of reasons, and is known as a *false negative.* Conversely, sensors occasionally produce output when no legitimate percept is present; this is a *false positive.*

Consider an infrared (IR) sensor. Suppose it is most sensitive at a wavelength of 1350 nm (nanometers). False negatives are least likely to occur for percepts at that wavelength. As the wavelength departs from 1350 nm, the sensitivity decreases, and the radiation must be more intense to cause the sensor to produce an output signal. The likelihood of false negatives increases as the wavelength becomes longer or shorter than 1350 nm. Outside a certain range of wavelengths, the sensor is relatively insensitive, and false negatives are therefore the rule rather than the exception. Whether the failure to produce output constitutes a false negative depends, however, on the range of wavelengths that are defined as "legitimate" percepts.

Suppose the sensor in the foregoing example is part of a proximity-detection device on a mobile robot. A laser on the robot, operating at a wavelength of 1350 nm, reflects from nearby objects in the work environment. The reflections are picked up by the sensor, which is covered by an IR filter that passes radiation easily within the range 1300 to 1400 nm, but blocks most energy outside that range. If the sensor output exceeds a certain level, the robot controller is instructed to change direction to avoid striking a possible obstruction. External sources of IR can cause false positives. This is most likely to occur if the external IR has a wavelength near the peak sensitivity region of the sensor/filter, that is, between 1300 and 1400 nm. However, if the external percept is sufficiently intense, it might cause a false positive even if its wavelength is considerably less than 1300 nm or greater than 1400 nm.

Robot controllers can be programmed to ignore false negatives or positives, as long as there is some way to distinguish between them and "legitimate" percepts. In a poorly designed system, however, false negatives or positives can cause erratic operation.

FAULT RESILIENCE

The term *fault resilience* can refer to either of two different characteristics of a computerized robotic system.

The first type of fault-resilient system can also be called *sabotage-proof*. Suppose that all the strategic (nuclear) defenses of the United States are placed under the control of a computer. It is imperative that it be impossible for unauthorized people to turn it off. Backup systems are necessary. No matter what anyone tries to do to cause the system to malfunction or become inoperative, the system must be capable of resisting or overcoming such attack.

Some engineers doubt that it is possible to build a totally sabotage-proof computer. They quote the saying, "Build a more crime-proof system, and you get smarter criminals." Also, any such system would have to be engineered and built by human beings. At least one of those people could be bribed or blackmailed into divulging information on how to defeat the security provisions. And of course, no one can anticipate all of the things that might go wrong with a system. According to Murphy's law, which is usually stated tongue-in-cheek but which can often manifest itself as truth, "If something can go wrong, it will." And the corollary, less often heard but perhaps just as true, is "If something cannot go wrong, it will."

The second type of fault resilience is also known as *graceful degradation*. Many computers and also computer-controlled robotic systems are designed so that if some parts fail, the system still works, although perhaps at reduced efficiency and speed. See GRACEFUL DEGRADATION.

FEEDBACK

Feedback is a means by which a *closed-loop system* regulates itself. Feedback is used extensively in robotics.

An example of feedback can be found in a simple thermostat mechanism, connected to a heating/cooling unit. Suppose the thermostat is set for 20 degrees Celsius (20°C). If the temperature rises much above 20°C, a signal is sent to the heating/cooling unit, telling it to cool the air in the room. If the temperature falls much below 20°C, a signal tells the unit to heat the room. This process is illustrated in the block diagram.

In a system that uses feedback to stabilize itself, there must be some leeway between the opposing functions. In the case of the thermostatically controlled heating/cooling system, if both thresholds are set for exactly 20°C, the system will constantly and rapidly cycle back and forth between

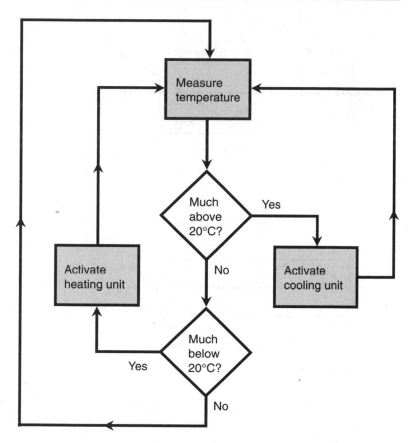

Feedback

heating and cooling. A typical range might be 18 to 22°C. The leeway should, however, not be too wide.

See also SERVOMECHANISM.

FIBER-OPTIC CABLE

A *fiber-optic cable* is a bundle of transparent, solid strands designed to carry modulated light or infrared (IR). This type of cable can carry millions of signals at high bandwidth.

Manufacture

Optical fibers are made from glass to which impurities have been added to maximize the transparency at certain wavelengths. The impurities also optimize the *refractive index* of the glass, or the extent to which it slows

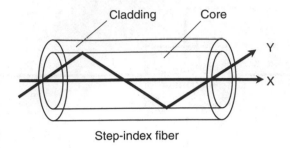

Step-index fiber

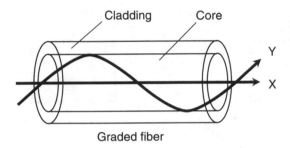

Graded fiber

Fiber-optic cable

down and bends light. An optical fiber has a *core* surrounded by a tubular *cladding*, as shown in the illustrations. The cladding has a lower refractive index than the core.

In a *step-index optical fiber* (top drawing), the core has a uniform index of refraction and the cladding has a lower index, also uniform. The transition at the boundary is abrupt. In the *graded-index optical fiber* (lower drawing), the core has a refractive index that is greatest along the central axis and steadily decreases outward from the center. At the boundary, there is an abrupt drop in the refractive index.

Operation

In the top illustration, showing a step-index fiber, ray X enters the core parallel to the fiber axis and travels without striking the boundary unless there is a bend in the fiber. If there is a bend, ray X veers off center and behaves like Y. Ray Y strikes the boundary repeatedly. Each time ray Y encounters the boundary, *total internal reflection* occurs, so ray Y stays within the core.

In the lower drawing, showing a graded-index fiber, ray X enters the core parallel to the fiber axis and travels without striking the boundary unless there is a bend in the fiber. If there is a bend, ray X veers off center and behaves like ray Y. As ray Y moves farther from the center of the core, the index of refraction decreases, bending the ray back toward the center. If ray Y enters at a sharp enough angle, it might strike the boundary, in which case total internal reflection occurs. Therefore, ray Y stays within the core.

Bundling

Optical fibers can be bundled into cable, in the same way as wires are bundled. The individual fibers are protected from damage by plastic jackets. Common coverings are polyethylene and polyurethane. Steel wires or other strong materials are often used to add strength to the cable. The whole bundle is encased in an outer jacket. This outer covering can be reinforced with wire mesh and/or coated with corrosion-resistant compounds.

Each fiber in the bundle can carry several rays of visible light and/or infrared (IR), each ray having a different wavelength. Each ray can in turn contain a large number of signals. Because the frequencies of visible light and IR are much higher than the frequencies of radio-frequency (RF) currents, the *bandwidth* of an optical/IR cable link can be far greater than that of any RF cable link. This allows much higher data speed.

FIELD OF VIEW (FOV)

The *field of view (FOV)* of a directional sensor is a quantitative expression of the angular range within which it reacts properly to stimuli, or *percepts*. The FOV is defined in terms of x and y (major and minor) angles, and applies primarily to unidirectional sensors (that is, devices intended to pick up energy from one direction). These angles can be defined as radial, relative to the axis at which the sensor is most responsive, or diametric (twice the radial value).

The horizontal FOV of a sensor takes the shape of a cone in three-dimensional (3-D) space, with the apex at the sensor, as shown in the illustration. This cone does not necessarily have the same flare angle in all planes passing through its axis. As "seen" from the point of view of the sensor itself, the cone appears as a circle or ellipse in an image of the work environment. If the FOV cone is circular, then the x and y angles are the same. If the FOV cone is elliptical, then the x and y angles differ. Compare RANGE.

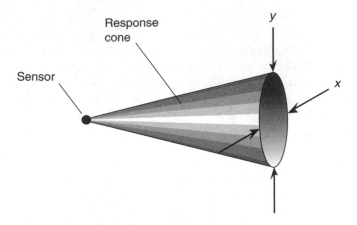

Response
cone

Sensor

y

x

Field of view (FOV)

FINE MOTION PLANNING

Fine motion planning refers to the scheme used by a robot to get into exactly the right position.

Suppose a personal robot is told to switch on the light in a hallway. The light switch is on the wall. The robot controller has a *computer map* of the house, and this includes the location of the hallway light switch. The robot proceeds to the general location of the switch, and reaches for the wall. How does it know exactly where to find the switch, and how to position its gripper precisely so that it will move the toggle on the switch?

One method is to incorporate robot vision, such as an *eye-in-hand system*. This allows the robot to recognize the shape of the toggle and guide itself accordingly. Another method involves the use of *tactile sensing*, so the *end effector* can "feel" along the wall in a manner similar to the way a human would find and actuate the switch with eyes closed. Yet another scheme might involve a highly precise, scaled-down *epipolar navigation* scheme. Compare GROSS MOTION PLANNING.

See also COMPUTER MAP, EPIPOLAR NAVIGATION, EYE-IN-HAND SYSTEM, TACTILE SENSING, and VISION SYSTEM.

FIRE-PROTECTION ROBOT

One role for which robots are especially well suited is fire fighting. If all fire fighters were robots, there would be no risk to human life in this occupation. Robots can be built to withstand far higher temperatures than humans can tolerate. Robots do not suffer from smoke inhalation. The main challenge is to program the robots to exercise judgment as keen as that of human beings, in a wide variety of situations.

One way to operate *fire-protection robots* is to have human operators at a remote point, and to equip the machines with *telepresence*. The operator sits at a set of controls, or wears a full-body suit with controls incorporated. When the operator moves a certain way, the robot moves in exactly the same way. Television cameras in the robot transmit images to the operator. The operator can "virtually" go where the robot goes, without any of the attendant risk.

One of the primary duties of household personal robots is to ensure the safety of the human occupants. This must include escorting people from the house if it catches fire, and then putting out the fire and/or calling the fire department. It might also involve performing some first-aid tasks.

See also TELEPRESENCE.

FIRMWARE

Firmware is a term referring to computer programs that are permanently installed in a system. Usually this is done in *read-only memory (ROM)*.

The firmware in a computer can be altered, but this requires a hardware change. This might mean physically replacing an *integrated circuit (IC)*, but there are devices whose firmware can be erased and then reprogrammed. These are called *erasable programmable read-only memory (EPROM)* ICs. Special equipment is needed to change the contents of an EPROM.

Firmware programming is common in microcomputer-controlled appliances and machinery, such as *fixed-sequence robots* that perform a given task repeatedly. Compare HARD WIRING.

FIXED-SEQUENCE ROBOT

A *fixed-sequence robot* is a robot that performs a single, preprogrammed task or set of tasks, making exactly the same movements each time. There is no exception or variation to the routine.

Fixed-sequence robots are ideally suited to assembly-line work. An entertaining example of a fixed-sequence robot is a toy that goes through some routine whenever a button is pressed. These machines are especially popular in Japan. In some cases, such toy robots appear sophisticated.

FLEXIBLE AUTOMATION

Flexible automation refers to the ability of a robot or system to do various tasks. To change from one task to another, a simple software change, or a change in the commands input to the controller, is all that is necessary.

A simple example of flexible automation is a robot arm that can be programmed to insert screws, drill holes, sand, weld, insert rivets, and spray paint on objects in an assembly line.

As personal robots evolve, they become capable of doing many things on the basis of a single, sophisticated program. This is the ultimate in flexible automation, and can be considered a form of *artificial intelligence (AI)*. The appropriate actions result from verbal commands. This necessitates speech-recognition capabilities, as well as considerable controller memory, speed, and processing power.

FLIGHT TELEROBOTIC SERVICER

In space missions, it is often necessary to perform repairs and general maintenance in and around the spacecraft. It is not always economical to have astronauts do this work. For this reason, various designs have been considered for a robot called a *flight telerobotic servicer (FTS)*.

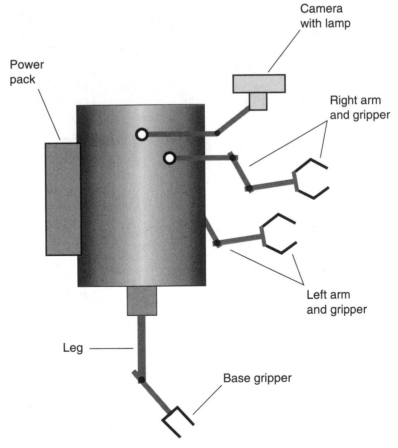

Flight telerobotic servicer

The FTS is a remote-controlled robot. The extent to which it is controlled depends on the design. The simplest FTS machines are programmable from the spacecraft's main computer. More complex FTS devices make use of *telepresence.*

Because of the risk involved in sending humans into space, scientists have considered the idea of launching FTS-piloted space shuttles to deploy or repair satellites. The FTSs would be controlled through computers on the ground and in the spacecraft. One FTS design has the appearance of a one-legged, headless *android,* as shown in the illustration.

See also TELEPRESENCE.

FLOWCHART

A *flowchart* is a diagram that illustrates a logical process or a computer program. It is a block diagram. Boxes indicate conditions, diamonds indicate decision points, and arrows show procedural steps.

Flowcharts are used to develop computer software. They are also used in troubleshooting of complex equipment. Flowcharts lend themselves well to robotic applications, because they indicate choices that a robot must make while it accomplishes a task.

A flowchart must always represent a complete process. There should be no places where a technician, computer, or robot will be left without some decision being made. There must be no *infinite loops,* where the process goes in endless circles without accomplishing anything.

Examples of flowcharts are shown in the definitions of BRANCHING and FEEDBACK.

FLUXGATE MAGNETOMETER

A *fluxgate magnetometer* is a computerized robot guidance system that uses magnetic fields to derive position and orientation data. The device uses coils to sense changes in the *geomagnetic field* (Earth's magnetic field), or in an artificially generated reference field.

Navigation within a defined area can be carried out by having the robot controller constantly analyze the orientation and intensity of the magnetic flux field generated by strategically placed electromagnets. A computer map of the flux field, showing two electromagnets and a hypothetical robot in the field, is shown in the illustration. In this case, opposite magnetic poles (north and south) face each other, giving the flux field a characteristic bar-magnet shape.

For each point in the work environment, the magnetic flux has a unique orientation and intensity. Therefore, there is a one-to-one correspondence between magnetic flux conditions and each point inside the environment. The robot controller is programmed to "know" this

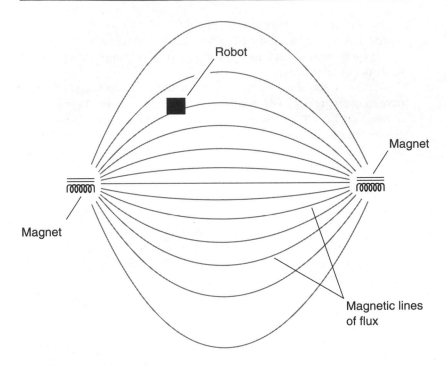

Robot

Magnet

Magnet

Magnetic lines
of flux

Fluxgate magnetometer

relation precisely for all points in the environment. This allows the robot to pinpoint its position in three-dimensional (3-D) space, provided a set of reference coordinates is established.

See also COMPUTER MAP.

FLYING EYEBALL

The *flying eyeball* is a simple form of *submarine robot*. This robot can resolve detail underwater, and can also move around. It cannot manipulate anything; it has no robot arms or *end effectors*. Flying eyeballs are used in scientific and military applications.

A cable, containing the robot in a special launcher housing, is dropped from a boat. When the launcher gets to the desired depth, it lets out the robot, which is connected to the launcher by a tether, as shown in the illustration. The tether and the drop cable convey data back to the boat. The robot contains a video camera and one or more lamps to illuminate the undersea environment. It also has a set of thrusters, or propellers, that let it move around according to control commands sent through the cable and tether. Human operators on board the boat can watch the images

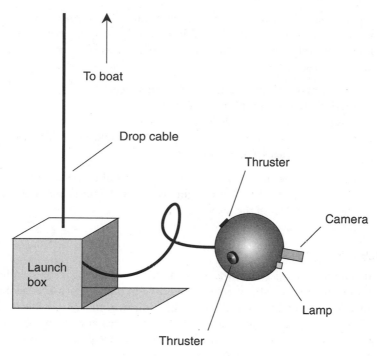

Flying eyeball

from the television camera, and guide the robot around as it examines objects on the sea floor.

In some cases, the tether can be eliminated, and radio-frequency (RF), infrared (IR), or visible beams can be used to convey data from the robot to the launcher. This allows the robot to have enhanced freedom of movement, without the concern that the tether might get tangled up in something. However, the range of the RF, IR, or link is limited, because water does not propagate these forms of energy for long distances.

FOOD-SERVICE ROBOT

Robots can be used to prepare and serve food. The major applications are in repetitive chores, such as placing measured portions on plates, assembly-line style, to serve a large number of people. *Food-service robots* are also used in canning and bottling plants, because these jobs are simple, repetitive, mundane, and easily programmable. As a row of bottles goes by, for example, one robot fills each bottle. Then a machine checks to be sure each bottle is filled to the right level. Rejects are thrown out by another robot. Still another robot places the caps on the bottles.

Personal robots, when they are programmed to prepare or serve food, require more autonomy than robots in large-volume food service. A household robot might be programmed to prepare a meal of meat, vegetables, and beverages. The robot would ask questions such as these:

- How many people will there be for this meal?
- Which type of meat is to be served?
- Which type of vegetable is to be served?
- How would you like the potatoes done? Or would you rather have rice?
- What beverages would you like?

When all the answers were received, the robot would carry out the task of preparing the meal. The robot might also serve the meal, and then clean up the table and wash the dishes afterwards.

See also **PERSONAL ROBOT.**

FORESHORTENING

In a robotic *distance-measurement* system, *foreshortening* is a false indication of the distance between a robot and a barrier, as measured along a specific straight-line path through three-dimensional (3-D) space. The phenomenon can occur when a barrier is oriented at a sharp angle with respect to the direction in which the range bearing is to be obtained. *Sonar* is particularly vulnerable to the problem, because it is difficult to focus acoustic waves into narrow beams.

The illustration shows a dimensionally reduced example of how foreshortening can take place. The robot is shown as a shaded circle at left. Its direction of travel, and the favored direction (axis) of its sonar device, is directly from left to right (horizontally in this drawing). The sonar should ideally produce a range indication that is the same as the *actual range,* or the distance the robot must travel before it runs into the barrier. However, the field of view (FOV) of the sonar is 30°, or 15° to either side of the axis. The extreme right-hand edge of the sonar beam strikes the barrier before the central portion of the beam. Assuming the barrier has a surface sufficiently irregular to scatter the acoustic waves in all directions so the robot receives an echo from all portions of its sonar beam, the *apparent range* is significantly less than the actual range.

The only solution to foreshortening problems of this sort is to minimize the FOV of the ranging equipment. In a work environment such as that shown in the drawing, the robot would be better off plotting a *computer map* of its surroundings, using a system more sophisticated than sonar.

See also **COMPUTER MAP, DISTANCE MEASUREMENT, FIELD OF VIEW (FOV),** and **RANGE PLOTTING.**

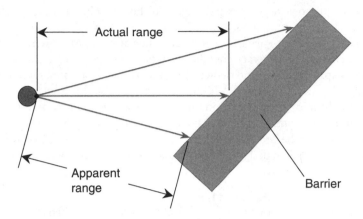

Actual range

Apparent range

Barrier

Foreshortening

FORWARD CHAINING

A computer can act as a person knowledgeable in some field, such as engineering, weather forecasting, medicine, or even the stock market. Programs that make computers act like specialists are called *expert systems*. When running an expert system, you supply the computer with information, and the computer solves a problem based on that information.

There are two ways in which the data can be supplied when using an expert system. You can input the facts one at a time, as the computer requests them; or you can input all the data at once, before the program begins working toward a solution. The latter method is *forward chaining*. The chain of reasoning starts from a single set of facts, and works forward until the problem is solved or a conclusion is reached.

After a computer receives the data in a forward-chaining expert system, the *inference engine* uses rules, written in the software, to infer a solution or conclusion. If more information is necessary, the computer will let the operator know, usually by asking specific questions. Compare BACK-WARD CHAINING.

See also EXPERT SYSTEM.

FRAME

A *frame* is a mental symbol, a means of representing a set of things. Frames can be envisioned as "windows in the mind." In *artificial intelligence (AI)*, objects and processes can be categorized in frames.

Suppose a robot is given the command, "Go to the kitchen and pour some water into a paper cup." The robot goes through a series of deductions concerning how to get this beverage, and how to obtain the object in which it is to be contained.

First, the robot goes to the kitchen. Then it begins a search for the particular kind of beverage container that has been specified, in this case a paper cup. The illustration depicts this process. The first frame represents all the objects in the kitchen. Within this frame, a subframe is selected: eating and drinking utensils. Within this, the appropriate frame contains cups and tumblers; within this frame, the desired category is paper cups. Even this subset can be broken down further. One might specify 12-oz paper cups, white in color, designed to withstand hot beverages as well as cold.

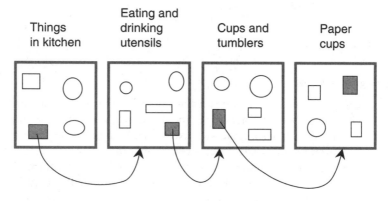

| Things in kitchen | Eating and drinking utensils | Cups and tumblers | Paper cups |

Frame

Frames can apply to procedures as well as to the selection of objects. Once the robot has the proper utensil in its grasp, what is to be done? Did the robot's user (human) want tap water, or is there some bottled water in the refrigerator? How about canned soda water? Maybe the user wants some of that mineral water she ran out of last week, in which case the robot must either come back and ask for further instructions, or else make a guess as to what substitute the user might accept.

FRANKENSTEIN SCENARIO

Science fiction is replete with stories in which some of the characters are robots or smart computers. Science-fiction robots are often *androids*. Such machines are invariably designed with the idea of helping humanity, although it often seems that the machines play roles in which some humans are "helped" at the expense of others.

A recurring theme in science fiction involves the consequences of robots, or intelligent machines, turning against their makers, or coming to logical conclusions intolerable to humanity. This theme is called the *Frankenstein scenario,* after the famous fictional android.

A vivid example of the Frankenstein scenario is provided by the novel *2001: A Space Odyssey,* in which Hal, an artificially intelligent computer on a space ship, tries to kill an astronaut. Hal somehow malfunctions, becomes paranoid, and believes that Dave, the astronaut, is intent on the computer's destruction. Ironically, Hal's paranoia brings about the very misfortune Hal dreads, because Dave is forced to disable Hal to save his own life.

A machine might react logically to preserve its own existence when humans try to "pull the plug." This could take the form of apparently hostile behavior, in which robot controllers collectively decide that humans must be eliminated. Because robots are supposed to preserve themselves according to *Asimov's three laws,* a robotic survival instinct can be useful, but only up to a certain point. A robot must never harm a human being; that is another of Asimov's laws.

Another example of the Frankenstein scenario is the team of computers in *Colossus: The Forbin Project.* In this case, the machines have the best interests of humanity in mind. War, the computers decide, cannot be allowed. Humans, the computers conclude, require structure in their lives, and must therefore have all their behavior strictly regulated. The result is a totalitarian state run by a machine.

See also ASIMOV'S THREE LAWS.

FRONT LIGHTING

In a robotic vision system, the term *front lighting* refers to illumination of objects in the work environment using a light source located at or near the robot's own imaging sensors. The light from the source therefore reflects from the surfaces of the objects under observation before reaching the sensors. Because the location of the lamp is near the sensors, the robot sees minimal, or no, shadow effect in its work environment.

Front lighting is used in situations where the surface details, particularly differences in color or shading, of observed objects are of interest or significance. For texture to show up, however, *side lighting* works best. Front lighting does not work particularly well in situations involving translucent or semitransparent objects, if their internal structure must be analyzed. *Back lighting* works best in these cases. Compare BACK LIGHTING and SIDE LIGHTING.

FUNCTION

A *function* is a *mapping* between set of objects or numbers. Functions are important in mathematics, and also in logic.

The drawing shows an example of a function as a mapping between two sets. Not all of the elements in the left-hand set (a few of which are shown by black dots) necessarily have counterparts in the right-hand set. Similarly, not all of the elements in the right-hand set (a few of which are shown by white dots) necessarily have counterparts in the left-hand set. If the mapping is to qualify as a function, it is possible for more than one element from the left-hand set to be mapped onto a single element in the right-hand set, but no element in the left-hand set can have more than one mate in the right-hand set. A function never maps a single element into more than one counterpart.

Set mapped from

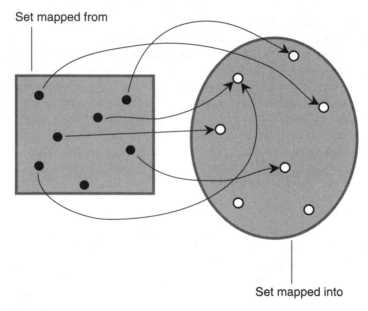

Set mapped into

Function

As shown in the illustration, the set of all elements on the left that have mates in the right is called the *domain* of the function. The *range* of the function is the set of all elements on the right with corresponding elements in the set on the left.

See also DOMAIN OF FUNCTION and RANGE OF FUNCTION.

In logic, a *function*, more specifically called a *logic function*, is an operation that takes one or more input variables, such as X, Y, and Z, and generates a specific output for each combination of inputs. Logic functions are generally simpler than mathematical ones, because the input variables can only have two values: 0 (false) or 1 (true).

An example of a logic function in three variables is shown in the table. First, the logic AND operation is performed on X and Y. Then the logic OR operation is performed between (X AND Y) and the variable Z. Some logic functions have dozens of input variables; there is only one output value, however, for each combination of inputs.

Function: example of a logic function

X	Y	Z	$f(X, Y, Z)$
0	0	0	0
0	0	1	1
0	1	0	0
0	1	1	1
1	0	0	0
1	0	1	1
1	1	0	1
1	1	1	1

Logic functions are important to engineers in the design of digital circuits, including computers. Often, there are several different possible combinations of logic gates that will generate a given logic function. The engineer's job is to find the simplest and most efficient design.

See also LOGIC GATE.

The term *function,* or more specifically *intended function,* is often used in reference to the set of tasks or routines that a robotic device or controller is designed to perform or execute. This definition is completely independent of the mathematical and logical definitions.

The mathematical equation, or set of equations, that represents a signal waveform is sometimes called a *function.* A *function generator* is a specialized circuit that generates waveforms whose curves are the graphs of specific mathematical functions. See GENERATOR.

FUNCTION GENERATOR

See GENERATOR.

FUTURIST

A *futurist* is a person who tries to predict, based on current technology and trends, what will be accomplished in a given scientific field in 5, 10, 50, 100, or more years. In robotics and *artificial intelligence (AI),* there is plenty of work for futurists.

Most futurists agree that robots will become more sophisticated, and more commonplace, as time goes by. There is some question as to exactly what form the robots will take. While it is fun to daydream about *androids,* these are often not the most practical and functional robots.

There is theoretically unlimited potential in AI. In practice, however, things have moved more slowly than futurists of the twentieth century hoped. Reasoning processes are incredibly complex. Some futurists believe that all human thought processes can be broken down into interactions among particles of matter. If this is true, then it is technically possible (although difficult) to build a computer that is as smart as, or smarter than, a human being. Other scientists are convinced that human thought involves factors that cannot be defined or replicated in purely material terms. If this is the case, then a computer with superhuman intelligence might be impossible to build.

Science-fiction authors have historically told stories about machines and scenarios, many of which have later become reality to a greater or lesser extent. For this reason, science-fiction writers have been called futurists.

G

GANTRY ROBOT

A *gantry robot* consists of a *robot arm* and *end effector* that employs three-dimensional (3-D) *Cartesian coordinate geometry* for precise positioning.

In one version of the gantry system, z-axis (up/down) movement is provided by a vertical shaft along which an assembly can slide. That assembly consists of a horizontal shaft, along which a horizontal arm at right angles to the shaft can slide on the *y* axis (forward/backward). A *cable drive* facilitates extension and retraction of the horizontal arm for *x*-axis (left/right) motion of the end effector.

Gantry robots are used in industrial robotics to position end effectors over specific points on a horizontal plane surface. The end effector can be a gripper that picks up or releases objects, as in *drop delivery*. Alternatively, a rotating-shaft end effector can be used, as in a robot designed to tighten bolts.

See also CABLE DRIVE, CARTESIAN COORDINATE GEOMETRY, END EFFECTOR, ROBOT ARM, *X* AXIS, *Y* AXIS, and *Z* AXIS.

GAS STATION ROBOT

Despite the rise in popularity of self-service gas stations, there are still people who would rather sit in their cars and have someone—or something—else do the dirty work. Robots are quite capable of filling your gas tank and washing your windshield.

The drawing illustrates what a typical robotized drive-through filling station, or *gas station robot,* might look like. A person drives a car up to the paying station and inserts a credit card. This card has information concerning the make and year of the car, as well as credit account data. This tells the robot where it can find the gas-tank fill opening (right or left side of the car), and whether there are enough funds in the credit account to pay for a full tank of gas. Another method of car identification might

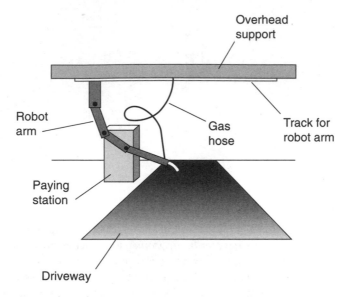

Overhead support

Robot arm

Gas hose

Track for robot arm

Paying station

Driveway

Gas station robot

be the use of *bar coding* or a *passive transponder,* similar to the price tags on consumer merchandise.

The robot must know the position of the car to within a millimeter or so. Otherwise, the nozzle might miss the fill pipe and spill gasoline on the pavement, or worse, put it in the car through the window. *Object recognition* helps prevent problems like this. Alternatively, a *biased search* can be used, letting the nozzle seek out the gas-tank fill opening. The opening itself is of a design such that the robot can open it and insert the nozzle, without any assistance from the human driving the car.

Robotized filling stations, should they become the norm, will let people stay in their cars without getting dirty, cold, wet, or hot. The service of a well-designed, robotized gas station should be fast and efficient. Robots will have to be programmed not to "top off" the gas tank to get a round number for the price. (This can result in overfilling the tank, and takes unnecessary time.) The robot will not forget to replace the gas cap, a perennial problem for some people who use "self-serve" gas stations.

GATEWAY

A *gateway* is a decision point in a specialized robotic-navigation process known as *topological path planning.* When a robot encounters a gateway, a decision must be made that affects the future path of the machine.

An example of a gateway is an intersection between two streets. At a typical intersection where two straight roads cross each other at a right angle, a robotized vehicle can do any of four things:

- Continue straight ahead
- Turn left
- Turn right
- Backtrack

When a mobile robot is programmed to travel from one point to another, gateways must frequently be dealt with. If the machine has a complete *computer map* of its work environment, and if the environment is not too complicated, every gateway possibility can reside in the controller's memory or storage medium. If the work environment is complex, or if it changes with time, the decisions must be based on programming rather than brute-force data storage.

See also **COMPUTER MAP, RELATIONAL GRAPH,** and **TOPOLOGICAL PATH PLANNING.**

GENERATOR

The term *generator* can refer to either of two devices. A *signal generator* is a source of alternating-current (AC) signal current, voltage, or power in an electronic circuit. An *oscillator* is a common example. An *electric generator* is a device that produces AC electricity from mechanical energy.

Signal generator

A signal generator is used for the purpose of testing audio-frequency (AF) or radio-frequency (RF) communications, detection, monitoring, security, navigation, and entertainment equipment. This includes various types of robotic sensing systems.

In its simplest form, a signal generator consists of a simple electronic oscillator that produces a *sine wave* of a certain amplitude in microvolts (μV) or millivolts (mV), and a certain frequency in hertz (Hz), kilohertz (kHz), megahertz (MHz), or gigahertz (GHz). Some AF signal generators can produce several different types of waveforms, such as those shown in Fig. 1. The more sophisticated signal generators for RF testing have *amplitude modulators* and/or *frequency modulators*.

A *function generator* is a signal generator that can produce specialized waveforms selected by the user. All electrical waveforms can be expressed as mathematical *functions* of time. For example, the instantaneous amplitude of a sine wave can be expressed in the form $f(t) = a \sin bt$, where a is a constant that determines the peak amplitude and b is a constant that determines the frequency. Square waves, sawtooth waves, and all other periodic

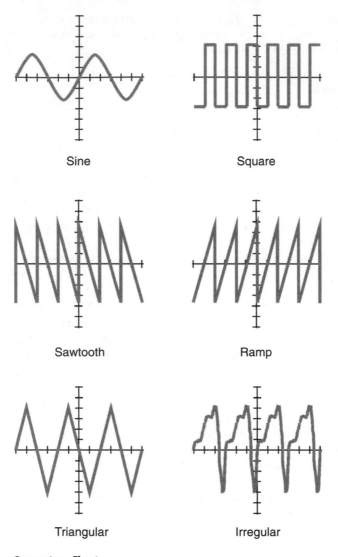

Sine   Square

Sawtooth   Ramp

Triangular   Irregular

Generator—Fig. 1

disturbances can be expressed as mathematical functions of time, although the functions are complicated in some cases.

Most function generators can produce sine waves, sawtooth waves, and square waves. Some can also produce sequences of pulses. More sophisticated function generators that can create a large variety of different waveforms are used for testing purposes in the design, troubleshooting, and alignment of electronic apparatus.

Electric generator

An electric generator is constructed somewhat like a conventional electric *motor*, although it functions in the opposite sense. Some generators can also operate as motors; they are called *motor/generators*. Generators, like motors, are energy transducers of a special sort.

A typical generator produces AC when a coil is rotated rapidly in a strong magnetic field. The magnetic field can be provided by a pair of permanent magnets (Fig. 2). The rotating shaft is driven by a gasoline-powered motor, a turbine, or some other source of mechanical energy. A *commutator* can be used with a generator to produce pulsating direct-current (DC) output, which can be filtered to obtain pure DC for use with precision equipment.

See also **MOTOR**.

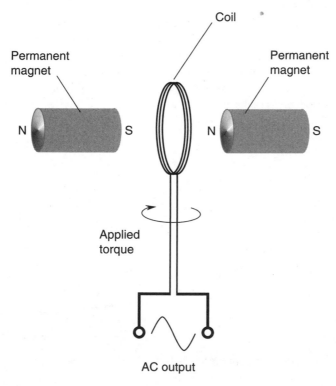

Generator—Fig. 2

GLOBAL POSITIONING SYSTEM (GPS)

The *Global Positioning System* (GPS) is a network of wireless location and navigation apparatus that operates on a worldwide basis. The GPS employs several satellites, and allows determination of latitude, longitude, and altitude. It is used in some mobile robotic systems for guidance when extreme, localized precision is not necessary.

All GPS satellites transmit signals in the ultra-high-frequency (UHF) radio spectrum. The signals are modulated with codes that contain timing information used by the receiving apparatus to make measurements. A GPS receiver determines its location by measuring the distances to four or more different satellites, and using a computer to process the information received from the satellites. From this information the receiver can give the user an indication of position accurate to within a few meters.

See also DISTANCE MEASUREMENT.

GRACEFUL DEGRADATION

When a portion of a computer system malfunctions, it is desirable to have the computer keep working even if the efficiency is impaired. If a single component causes the whole computer to fail, it is called a *catastrophic failure*. This can generally be prevented by good engineering, including the use of backup systems. In *graceful degradation,* as the number of component failures increases, the efficiency and/or speed of the system gradually

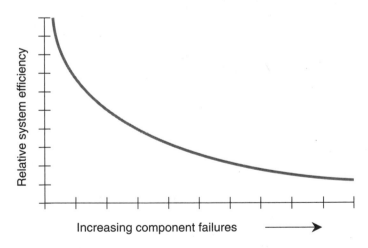

Graceful degradation

declines, but does not instantly drop to zero. The illustration is a graph of the behavior of a hypothetical robotic system with graceful degradation.

In the event of a subsystem malfunction, a sophisticated computer or robot controller can use other circuits to accomplish the tasks of the failed part of the system temporarily. The human operator or attendant is notified that something is wrong, and technicians can fix it, often with little or no downtime. Compare FAULT RESILIENCE.

GRAPHICAL PATH PLANNING

Graphical path planning is a method of navigation used by *mobile robots.* It is a specialized scheme or set of schemes for the execution of *metric path planning.* In graphical path planning, all possible routes are plotted on a *computer map* of the work environment. These routes can be chosen in various ways, by employing specific algorithms.

In an open work environment (that is, one in which there are no hazards or obstructions), the best routes are usually straight lines between the *nodes,* or stopping points (Fig. 1). The algorithm for determining these paths is comparatively simple; it can be represented by a set of linear equations in the robot controller. An obstacle, barrier, or hazard can complicate this scenario, but only if it intersects, or nearly intersects, one of the lines determined by the linear equations. To avoid mishaps, the algorithm can be modified to include a statement to the effect that the machine must never come closer than a certain distance to an obstacle, barrier, or hazard. *Proximity sensing* can be employed to detect these situations.

In a work environment in which there are numerous obstacles or hazards, or where there are barriers such as walls separating rooms and hallways, the straight-line algorithm is not satisfactory, even in amended form, because too many modifications are necessary. One scheme that works quite well in this type of environment is the *Voronoi graph.* The

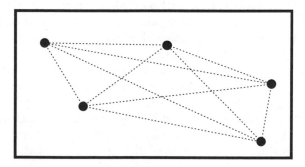

Graphical path planning—Fig. 1

paths are defined as sets of points at the greatest possible distances from obstacles, barriers, or hazards. In a hallway, for example, the path goes down the middle. The same is true as the robot passes through doorways. The paths in other places depend on the locations of the nodes, and the arrangement of obstructions in the rooms or open areas (Fig. 2).

See also COMPUTER MAP. Compare METRIC PATH PLANNING and TOPOLOGICAL PATH PLANNING.

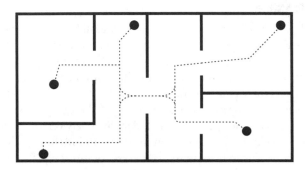

Graphical path planning—Fig. 2

GRASPING PLANNING

Grasping planning refers to the scheme that a robot arm and gripper use to get hold of a chosen object.

Suppose a person tells a robot to go to the kitchen and get a spoon. The robot uses *gross motion planning* to find the kitchen, and *fine motion planning* to locate the correct drawer and determine which objects in the drawer are the spoons. Then the gripper must grasp a spoon, preferably by the handle rather than by the eating end. The robot must not get a fork, or two spoons, or a spoon along with something else such as a can opener.

Hopefully, the silverware is arranged logically in the drawer, so spoons are not randomly mixed up with forks, knives, can openers, and other utensils. This can be ensured by programming, as long as the robot (but only the robot) has access to the drawer. If there are children in the household, and if they get into the silverware drawer, the robot had better be able to cope with mixed-up utensils. Then, getting a spoon becomes a form of *bin picking problem*.

Close-up, detailed machine vision, such as an *eye-in-hand system*, can ensure that the gripper gets the right utensil in the right way. *Tactile sensing* might also be used, because a spoon "feels" different than any other kind of utensil.

See also BIN PICKING PROBLEM, EYE-IN-HAND SYSTEM, FINE MOTION PLANNING, GROSS MOTION PLANNING, OBJECT RECOGNITION, and TACTILE SENSING.

GRAVITY LOADING

Gravity loading is a phenomenon that introduces positioning error into robot arms as a result of the force of gravity.

All robot arms are comprised of materials that bend or stretch to some extent; no known substance is perfectly rigid. In addition, all materials have some mass; thus, in a gravitational field they also have weight. The weight of the robot arm and end effector always causes some bending and/or stretching of the materials from which the assembly is made. The effect can be exceedingly small, as in a telescoping, vertically oriented robot arm; or it can be larger, as in a long, jointed robotic arm. However, the effect is never entirely absent.

The error caused by gravity loading is not always significant. In situations where gravity loading causes significant positioning errors, a scheme for correction is necessary.

See also ERROR CORRECTION.

GRAYSCALE

Grayscale is a method of creating and displaying digital video images. As its name implies, a grayscale vision system is color-blind.

Each image is made up of *pixels*. One pixel is a single picture (pix) element. The pixels are tiny squares, each with a shade of gray that is assigned a digital code. There are three commonly used schemes for rendering pixels in grayscale: percentage-of-black, 16 shades of gray, and 256 shades of gray.

In the percentage scheme, there are usually 11 levels according to the following sequence: {black, 90 percent black, 80 percent black, ..., 20 percent black, 10 percent black, white}. Sometimes the brightness is broken down further, into increments of 5 percent or even 1 percent rather than 10 percent; such gradations tend to be imprecise because computer digital codes are binary (power-of-2), not decimal (power-of-10).

In the 16-shade scheme, four binary digits, or *bits*, are needed to represent each level of brightness from black = 0000 to white = 1111. In the 256-shade scheme, eight binary digits are used, from black = 00000000 to white = 11111111.

See also COLOR SENSING and VISION SYSTEM.

GRIPPER

See ROBOT GRIPPER.

GROSS MOTION PLANNING

Gross motion planning is the scheme a mobile robot employs to navigate in its work environment without bumping into things, falling down stairs, or tipping over. The term can also refer to the general, programmed sequence of movements that a robot arm undergoes in an industrial robotic system.

Gross motion planning can be done using a *computer map* of the environment. This tells it where tables, chairs, furniture, and other obstructions are located, and how they are oriented. Another method is to use *proximity sensing* or a *vision system*. These devices can work in environments unfamiliar to a robot, and for which it has no computer map. Still another method is the use of *beacons*.

Suppose a personal robot is told to go to the kitchen and get an apple from a basket on a table. The robot can employ gross motion planning to scan its computer map and locate the kitchen. Within the kitchen, it needs some way to determine where the table is located. Finding the basket, and picking an apple from it (especially if there are other types of fruit in the basket, too), requires *fine motion planning*. Compare FINE MOTION PLAN-NING and GRASPING PLANNING.

GROUNDSKEEPING ROBOT

There are plenty of jobs for personal robots in the yard around the house, as well as inside the house. Two obvious applications for a *groundskeeping robot* includes mowing the lawn and removing snow. In addition, such a machine might water and weed a garden.

Riding mowers and riding snow blowers are easy for sophisticated mobile robots to use. The robot need not be a biped; it needs only to have a form suitable for riding the machine and operating the controls. Alternatively, lawn mowers or snow blowers can be robotic devices, designed with that one task in mind.

The main challenge, once a lawn-mowing or snow-blowing robot has begun its work, is to do its work everywhere it is supposed to, but nowhere else. A robot owner does not want the lawn mower in the garden, and there is no point in blowing snow from the lawn (usually). Such a robot should therefore be an *automated guided vehicle* (AGV). Current-carrying wires can be buried around the perimeter of your yard, and along the edges of the driveway and walkways, establishing the boundaries within which the robot must work.

Inside the work area, *edge detection* can be used to follow the line between mown and unmown grass, or between cleared and uncleared pavement. This line is easily discernible because of differences in brightness and/or color. Alternatively, a *computer map* can be used, and the robot can sweep along controlled and programmed strips with mathematical precision.

The hardware already exists for groundskeeping robots to withstand all temperatures commonly encountered in both summer and winter, from Alaska to Death Valley. Software is more than sophisticated enough for ordinary yard-maintenance and snow-removal tasks. The only challenge remaining is to bring the cost down to the point that the average consumer can afford the robot.

See also AUTOMATED GUIDED VEHICLE, COMPUTER MAP, EDGE DETECTION, and PERSONAL ROBOT.

GUIDANCE SYSTEM

In robotics, *guidance system* refers to the hardware and software that lets a robot find its way in its work environment. In particular, it refers to *gross motion.* For detailed information, see AUTOMATED GUIDED VEHICLE, BEACON, BIASED SEARCH, COMPUTER MAP, DIRECTION FINDING, DIRECTION RESOLUTION, DISTANCE MEASUREMENT, DISTANCE RESOLUTION, EDGE DETECTION, EMBEDDED PATH, EPIPOLAR NAVIGATION, GLOBAL POSITIONING SYSTEM (GPS), GROSS MOTION PLANNING, GYROSCOPE, LOG POLAR NAVIGATION, OBJECT RECOGNITION, PARALLAX, PROXIMITY SENSING, RADAR, SONAR, and VISION SYSTEM.

GYROSCOPE

A *gyroscope* or *gyro* is a device that is useful in robot navigation. It forms the heart of an *inertial guidance system,* operating on the basis of the fact that a rotating, heavy disk tends to maintain its orientation in space.

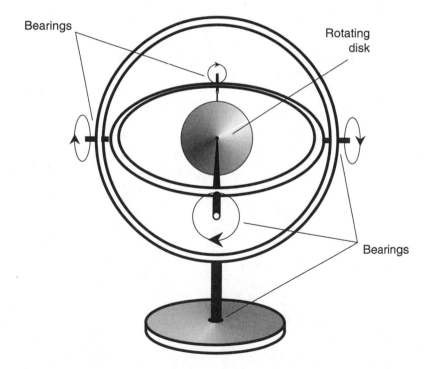

Gyroscope

The illustration shows the construction of a simple gyroscope. The disk, made of massive material such as solid steel or tungsten, is mounted in a *gimbal,* which is a set of bearings that allows the disk to turn up and down or from side to side; conversely, the bearings allow the entire assembly (except for the disk) to undergo *pitch, roll,* and *yaw* while the disk remains fixed in its spatial orientation. The disk is usually driven by an electric motor.

A gyroscope can be employed to keep track of a robot's direction of travel, or bearing, in three-dimensional (3-D) space without reliance on external objects, beacons, or force fields. Gyroscopes allow the accurate operation of guidance systems for a limited time, because they tend to change their orientation slowly over long periods. In addition, gyroscopes are susceptible to misalignment in the event of physical shock.

See also PITCH, ROLL, and YAW.

H

"HACKER" PROGRAM

One of the earliest experiments with artificial intelligence (AI) was done with an imaginary robot, entirely contained within the "mind" of a computer. A student named *Gerry Sussman* wrote a program called "*Hacker*," in a computer language known as *LISP*. The result was a little universe in which a robot could stack blocks on each other.

Sussman created laws of physics in the imaginary universe. Among them were things such as

- Blocks X, Y, and Z each weigh 5 lb.
- Blocks V and W each weigh 50 lb.
- The robot can lift no more than 10 lb.
- Only one object can occupy a given space at a given time.
- The robot knows how many blocks there are.
- The robot can find blocks if they are not in direct sight.

Illustration 1 shows the five blocks lying around, as they might appear on the computer monitor, along with the robot.

Sussman gave commands to the robot, such as, "Stack the blocks all up, one on top of the other." As stated, this command is impossible, because it requires the robot to lift a block weighing 50 lb (either V or W), and the robot is capable of lifting only 10 lb (see illustration 2). What would happen? Would the robot try forever to lift a block beyond its limit of strength? Or would it tell Sussman something like, "Unable to do this"? Would it go after either block V or W first, trying to get it on top of one of the lighter blocks, or on top of the other heavy block? Would it pick up all the lighter blocks X, Y, and Z in some sequence, stacking them vertically on top of V or W? Would it put two light blocks on V, and the remaining light block on W, and then give up? Eventually, the robot would run into the impossibility of the command. But how long would it try, and what would it try, before quitting?

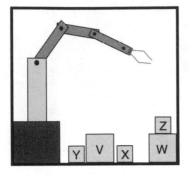

1. Why is X trying to hide?

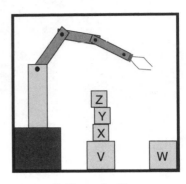

2. Now what?

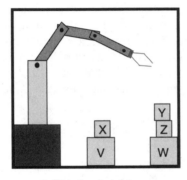

3. This is all right.

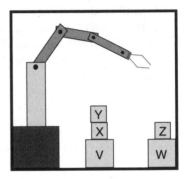

4. Is this all right too?

"Hacker" program

Another command might be, "Stack the blocks so that lighter ones are on top of heavier ones." This can be done according to the rules written above. But there exist several different possible ways (two of these are shown in illustrations 3 and 4). Would the robot hesitate, unable to make a decision? Or would it go ahead and accomplish the task in some way? If the experiment were repeated, would the result always be the same, or would the robot solve the problem a different way each time?

Numerous AI researchers have written programs similar to "Hacker," creating "computer universes" in an attempt to get machines to think and learn. The results have often been fascinating and unexpected.

HALLUCINATION

In a human being, a *hallucination* occurs when the senses deliver phantom messages. This can happen in mental illness, or under the influence of certain drugs. Hallucinations can be, and often are, combined with delusions, or misinterpretations of reality. An example is the person who thinks that spies are after him or her, and who sees sinister figures lurking behind trees or in dark alleyways.

Sophisticated computers can appear to have hallucinations and delusions. The likelihood of such malfunctions, taking place in bizarre and often inexplicable ways, increases as systems become more complex. This is because, as computers become smarter, the number of components, pathways, and nodes increases in exponential proportion, and the probability of a component failure or stray signal thus "blows up." Computer components are, in general, exceptionally reliable; however, given great enough numbers of them, strange things can happen, and have happened. Experienced personal computer users and technicians know this.

Some researchers in artificial intelligence (AI) believe that electronic hallucinations or delusions might someday result from improper design and care of machines. These researchers suspect that machines, as they evolve and become more intelligent, can develop "hangups," just as people do. At present, malicious human operators cause more problems directly, by means of such schemes as hacking and the writing of computer viruses, than "computers gone mad." In a few decades, however, *autonomous robots* might become able to program and maintain themselves to a large extent, and the situation might change. See AUTONOMOUS ROBOT.

HAND

See ROBOT GRIPPER.

HANDSHAKING

In a digital communications system, accuracy can be optimized by having the receiver verify that it has received the data correctly. This is done periodically—say, every three characters—by means of a process called *handshaking*.

The process goes as follows, as illustrated in the figure. First, the transmitter sends three characters of data. Then it pauses, and awaits a signal from the receiver that says either of the following:

- (*a*) All three characters have familiar formats.
- (*b*) One or more characters has an unfamiliar format.

If the return signal is (*a*), the transmitter sends the next three characters. If the return signal is (*b*), the transmitter repeats the three characters.

In computer systems, the term *handshaking* refers to a method of controlling, or synchronizing, the flow of *serial data* between or among devices. The synchronization is accomplished by means of a control wire in

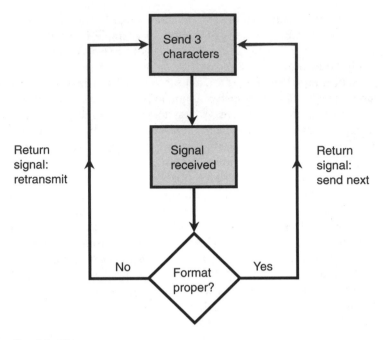

Handshaking

hardware, or a control code in the programming. *Hardware handshaking* is used when direct wire or cable links are possible, such as between a personal computer and a serial printer. *Software handshaking* is similar to the process used for communications systems.

HARD WIRING

In a computer or *autonomous robot,* the term *hard wiring* refers to functions that are built directly into the machine hardware. Hard wiring cannot be changed without rearranging physical components, or changing the interconnecting wires. Sometimes, the expression *firmware* is used to mean hard wiring, although technically that is a misuse of the term.

An ideal computer (that is, a computer with infinite processing power and zero error rate, which can exist only in theory) could be programmed to do anything without having to move a single physical component. Of course, the components must be hooked up together somehow, but in the ideal case, functions could be changed just by reprogramming the machine. This has been realized to a large extent in recent years by the use of high-speed, high-capacity data storage media.

Hard wiring does have some advantages over software control. Most significant is the fact that hard-wired functions can be done at a higher rate of speed than processes that require access to mechanical storage media. However, as nonvolatile storage media without moving parts become more widely available, this advantage of hard wiring will gradually erode. Compare FIRMWARE.

HERTZ

Hertz, abbreviated Hz, is the fundamental measure of alternating-current (AC) frequency. A frequency of 1 Hz is equivalent to one cycle per second. In fact, the word "hertz" is interchangeable with the expression "cycles per second."

Frequency is often expressed in units of *kilohertz (kHz), megahertz (MHz),* and *gigahertz (GHz).* A frequency of 1 kHz is equal to 1000 Hz; a frequency of 1 MHz is equal to 1000 kHz or 10^6 Hz; a frequency of 1 GHz is equal to 1000 MHz or 10^9 Hz.

The speed at which digital computers operate is often specified in terms of frequency. The higher the frequency, the faster a microprocessor can work, and the more powerful can be the computer that uses the chip— if all other factors remain constant. The reason that higher frequency translates into a more powerful chip is simply that, as the frequency increases, more and more instructions can be executed, and thus more operations done, per unit time. The *clock frequency* of the microprocessor is, however, only one of several factors that determine the processing speed of a computer.

HEURISTIC KNOWLEDGE

Can computers and robots learn from their mistakes, and improve their knowledge by trial and error? Is it possible for a machine, or a network of machines, to evolve on its own? Some *artificial intelligence (AI)* researchers believe so. The existence of *heuristic knowledge,* or the ability of a machine to become smarter based on its real-world experience— literally learning from its own mistakes—is a classical characteristic of true AI.

Suppose a powerful computer is developed that can evolve to higher and higher levels of knowledge. Imagine that, one day after the machine has been put into operation, it has intelligence equivalent to that of a 10-year-old human; and after two days, it is as smart (in a rudimentary sense) as a 20-year-old. Suppose that after three days, the machine has knowledge equivalent to that of a 30-year-old research engineer. Suppose that more and more memory is added, so that the limit of knowledge is determined only by the speed of the microprocessor. What will such a computer be like after a month? Will it have the knowledge of a 300-year-old person (if people lived that long)? Moreover, does an ever-increasing level of intelligence imply that a machine can also become "wise"?

Machine knowledge becomes far more powerful when computers are given the ability to control mechanical devices, as is the case with autonomous robots. Intelligence and knowledge alone cannot build cars, bridges, aircraft, and rockets. Perhaps dolphins are as smart as people, but these marine mammals lack hands and fingers with which to manipulate things. A computerized robot is to a computer as a human being is to a dolphin.

Can computers ever become smarter than, and perhaps more powerful than, their makers? Some scientists are concerned that AI will be misused, or that it could evolve on its own with unintended, unexpected, and unpleasant results. Other researchers believe that the potential benefits of ever-increasing machine knowledge will always outweigh the potential dangers, and that we can always pull the plug if things get out of control.

HEXADECIMAL NUMBER SYSTEM

See NUMERATION.

HIERARCHICAL PARADIGM

The term *hierarchical paradigm* refers to the oldest of three major approaches to robot programming. A robot that employs the hierarchical paradigm relies largely on advance planning to carry out its assigned tasks. In the most sophisticated robot systems, there are three basic functions, known as *plan/sense/act*. The hierarchical paradigm simplifies this to *plan/act*.

The original idea for this paradigm was based on an attempt to get a smart robot to mimic human thought processes. The robot first senses the nature of its work environment, plans an action or sequence of actions, and then carries out those actions. In some systems this process occurs only once, at the beginning of the task; in other systems the planning step is repeated at intervals during the execution of the task.

The hierarchical paradigm has also been called the *deliberative paradigm*, because of its reliance on creating fixed models of the work environment. The robot controller functions in a sense as if it is "cogitating" or "deliberating" a strategy prior to carrying it out. This scheme has proven too simplistic for many practical scenarios, and around the year 1990, it was superseded by more advanced programming methods. Compare HYBRID DELIBERATIVE/REACTIVE PARADIGM and REACTIVE PARADIGM.

HIGH-LEVEL LANGUAGE

The term *high-level language* refers to programming languages used by humans in their interactions with computers. The various high-level languages each have advantages in some types of work, and shortcomings in others.

High-level language consists of statements in English (or some other written human language). This allows people to work with computers on a sort of conversational level. Most students find high-level languages easy to learn. The best way to learn these languages is to "play computer," thinking strictly by rules of logic. Because of the pure logic in programming, computers might someday be used to develop new programs for other computers. Compare MACHINE LANGUAGE.

See also HEURISTIC KNOWLEDGE.

HOBBY ROBOT

A *hobby robot* is a robot intended mainly for amusement and experimentation. Such a machine is usually autonomous, and contains its own controller. It is, in effect, a sophisticated toy.

Hobby robots often take humanoid form; these are *androids*. They can be programmed to give lectures, operate elevators, and even play musical instruments. *Wheel drives* are commonly used rather than bipedal (two-legged) designs, because wheels work better than legs, are easier to design, and cost less. However, some hobby robots are propelled by *track drives*; others have four or six legs.

Some hobby robots are adaptations of industrial robots. *Robot arms* can be attached to a main body. *Vision systems* can be installed in the robot's head, which can be equipped to turn to the right and left, and to nod up and down. *Speech recognition* and *speech synthesis* can allow a hobby robot to converse with its owner in plain language, rather than by means of a keyboard and monitor. This makes the machine much more human-like and user-friendly.

Perhaps the most important feature for a hobby robot is *artificial intelligence (AI)*. The "smarter" the robot, the more fun it is to have around. It is especially interesting if a machine can learn from its mistakes, or be taught things by its owner.

Hobby robot societies exist in the United States and several other developed countries. They evolve and change their names often. If you live in a large city, you might be near such an organization.

See also PERSONAL ROBOT.

HOLD

Hold, also called *holding,* is a condition in which the movements of part, or all, of a robot manipulator are temporarily brought to a halt. When this occurs, braking power is maintained, so the halted parts resist movement if outside pressure is applied. Common methods of ensuring braking force involve the use of a *hydraulic drive* or a *stepper motor.*

Holding can be a part of the programmed movement sequence for a robot arm and *end effector*. A good example is a situation in which a *gantry robot* is used to position a component for *drop delivery*.

See also DROP DELIVERY, GANTRY ROBOT, HYDRAULIC DRIVE, and STEPPER MOTOR.

HOME POSITION

In a robot manipulator, the *home position* is a point at which the *end effector* normally comes to rest. When the robot is shut down, or when it must be reset, the machine reverts to its home position.

When a coordinate system is used to define the location of the end effector, the home position is often assigned to the origin point. Thus, for example, in a robot arm and end effector using two-dimensional (2-D) Cartesian coordinate geometry, the home position can be assigned the value $(x, y) = (0, 0)$.

HOUSEHOLD ROBOT

See PERSONAL ROBOT.

HUMAN ENGINEERING

Human engineering refers to the art of making machines, especially computers and robots, easy to use. This is sometimes also called *user-friendliness*.

A user-friendly computer program allows the machine to be operated by someone who knows nothing about computers. Bank automatic-teller machines (ATMs) are a good example of devices that employ user-friendly programming. Increasingly, libraries are computerizing their card catalogs, and it is important that the programs be user-friendly so that people can find the books they want. There are many other examples.

A user-friendly robot can carry out orders efficiently, reliably, and reasonably fast. Ideally, a human operator can say something like, "Go to the kitchen and get me an apple," and (assuming there are any apples in the kitchen) the robot will return in a minute or two, holding an apple. A seemingly simple task like this is difficult to program into a machine, as researchers have found out. Even the most basic tasks are complex in terms of the number and combination of digital logic operations.

One of the most important considerations in human engineering is *artificial intelligence (AI)*. It is much easier to communicate with a machine that is "smart," compared with one that is "stupid." It is especially enjoyable if the machine can learn from its mistakes, or show ability to reason. *Speech recognition* and *speech synthesis* also help make computers and robots user-friendly.

See also HEURISTIC KNOWLEDGE, SPEECH RECOGNITION, and SPEECH SYNTHESIS.

HUMANOID ROBOT

See **ANDROID.**

HUNTING

Hunting is the result of overcompensation in a *servomechanism.* It is especially likely when there is not enough *hysteresis,* or sluggishness, in the system response.

Any circuit or device designed to lock onto something, by means of error correction, is subject to hunting. It takes the form of a back-and-forth oscillation between two conditions. If severe, it can go on indefinitely. In less serious cases, the system eventually settles on the correct level or position (see the illustration).

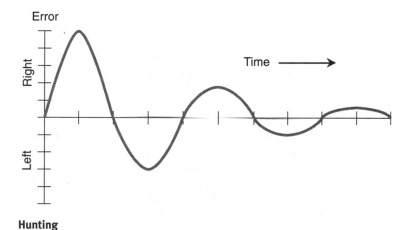

Hunting

Hunting is eliminated by careful design of feedback systems, so there is just the right amount of hysteresis. See **HYSTERESIS LOOP** and **SERVOMECHANISM.**

HYBRID DELIBERATIVE/REACTIVE PARADIGM

The *hybrid deliberative/reactive paradigm* is an approach to smart-robot programming that combines the attributes of two simpler schemes, known as the *hierarchical paradigm* and the *reactive paradigm.* The hybrid paradigm came into favor during the 1990s. It operates according to the principle *plan/sense/act.* The actions are based on advance planning and also on the outputs of sensors from moment to moment.

Before the task in begun, the robot generates a work plan. This is known as *mission planning,* and is a form of *deliberation.* A complex task is broken down into several components, or subtasks. Each subtask has its own subplan. Once the robot has begun executing the job, it carries out the plan and the subplans subject to modifications that may be necessary as the work environment changes. These changes are the results of signals from the sensors.

In a typical robot that uses the hybrid paradigm, deliberations occur at intervals of several seconds, while reactions take place at a rate of many times per second. Compare HIERARCHICAL PARADIGM and REACTIVE PARADIGM.

HYDRAULIC DRIVE

A *hydraulic drive* is a method of providing movement to a robot manipulator. It uses a special *hydraulic fluid,* usually oil-based, to transfer forces to various joints, telescoping sections, and *end effectors.*

The hydraulic drive consists of a power supply, one or more motors, a set of pistons and valves, and a feedback loop. The valves and pistons control the movement of the hydraulic fluid. Because the hydraulic fluid is practically incompressible, it is possible to generate large mechanical forces over small surface areas, or, conversely, to position large-area pistons with extreme accuracy. The feedback loop consists of one or more force sensors that provide error correction and ensure that the manipulator follows its intended path.

Hydraulically driven manipulators are used when motions must be rapid, precise, and repeated numerous times. Hydraulic systems are also noted for the ability to impart considerable force, so they are good for applications involving heavy lifting or the application of large amounts of pressure or torque. In addition, hydraulically driven robot manipulators resist unwanted movement in the presence of external forces. Compare PNEUMATIC DRIVE.

HYSTERESIS LOOP

A *hysteresis loop* (the word is pronounced "his-ta-REE-sis") is a graph that shows the sluggishness of response in a servomechanism.

The illustration shows a hysteresis loop for a typical thermostat, used for control of the indoor air temperature in a house. The horizontal scale shows the room temperature in degrees Celsius (°C). On/off conditions for heating and cooling are shown on the vertical scales. Notice that there is a small range of temperatures, from about 18.5°C to 21.5°C, within which the temperature fluctuates. This prevents the system from rapidly oscillating back and forth between heating and cooling states, but it is a narrow enough temperature range so that the people in the room don't get too hot or cold.

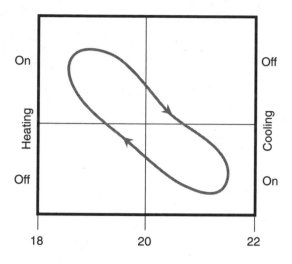

Temperature, Degrees Celsius

Hysteresis loop

All servomechanisms employ *feedback* of some kind. There must always be some hysteresis built into the feedback response. This hysteresis is often a natural result of the environment; for example, it takes some time for the temperature in a house to warm up and cool down by 3°C as depicted in the illustration. However, in a tiny temperature-regulated chamber intended to ensure the stable operation of the controller in a robot working in an extreme environment such as outer space, the hysteresis must be incorporated into the electronic design of the feedback circuit or thermostat. Otherwise, overreactions can be so severe that the system constantly cycles between states.

See also SERVOMECHANISM.

I

IF/THEN/ELSE

In computers and smart robots, choices must often be made in the execution of a program. One of the most common programming decision processes is called *IF/THEN/ELSE*. It can be expressed as a sentence: "If A, then B; otherwise (or else) C."

An example of an IF/THEN/ELSE process is shown in the illustration. The intent is to determine the absolute value of a real number. Suppose a computer is working with an input number, designated x. If x is negative (that is, if $x < 0$), then x must be multiplied by -1 to obtain the absolute value $|x|$. If x is zero or positive, then x is equal to its absolute value. The computer must compare the numerical value of x with zero. The machine will then output the absolute value of the number, by either multiplying x by -1 or by leaving x alone.

IF/THEN/ELSE processes are especially useful command structures for robots. You might tell a robot, "Go to the kitchen and get me a paper napkin." The robot controller has a command structure stored on its hard drive or in memory. It needs an alternative in case there are no paper napkins in the kitchen. The programming might take the form: "If this command can be executed, then carry out the task. Otherwise, output the audio statement, 'Your order cannot be completed because there are no paper napkins in the kitchen.'"

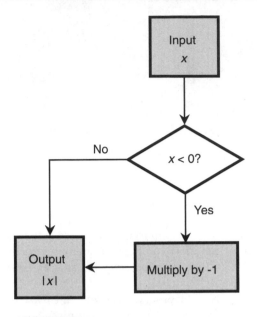

IF/THEN/ELSE

IGNORANT COEXISTENCE

See **COEXISTENCE.**

IMAGE ORTHICON

An *image orthicon* is a video camera tube, similar to a *vidicon*. It is useful in moderate and dim light.

The illustration is a simplified functional diagram of an image orthicon. A narrow beam of electrons, emitted from an *electron gun,* scans a target electrode. Some of this beam is reflected. The instantaneous amount of reflected electron-beam energy depends on the emission of *secondary electrons* from the target electrode. The number of secondary electrons depends on how much light is hitting the target electrode in a given place. The greatest return beam intensity corresponds to the brightest parts of the video image. The return beam is therefore amplitude-modulated as it scans the target electrode in a pattern that follows the scanning pattern in a television (TV) picture tube. The return beam strikes a sensor called a *photoreceptor.* The output of the photoreceptor is fed to an amplifier such as a photomultiplier. From there, the output is processed by a computer or robot controller.

The main limitation of the image orthicon is that it produces significant noise in addition to the signal output. However, when a fast response is

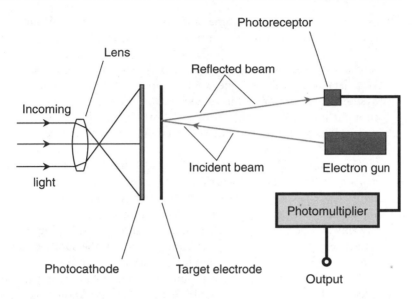

Image orthicon

needed (for example, when there is a lot of action in a scene) and the illumination intensity varies over a wide range, the image orthicon is useful. It can be employed in robot vision systems that process images rapidly, and/or that operate in work environments where the ambient light intensity can change dramatically. Compare CHARGE-COUPLED DEVICE and VIDICON.

See also VISION SYSTEM.

IMAGE RECOGNITION

See OBJECT RECOGNITION.

IMAGE RESOLUTION

See RESOLUTION.

IMMORTAL KNOWLEDGE

In advanced and developing nations, computers have brought about a transformation in human culture. The only role humans need to play, in the accumulation of knowledge in the general electronic database, is to input data into the systems.

Before computers existed (that is, before about 1950), history was passed from generation to generation in the form of books and verbal stories. If you read a book written 200 years ago, you interpret events somewhat differently than the original author thought of them. This is because society is not the same as it was two centuries ago. Values have changed. People have different priorities and beliefs.

When history is put down in books, or told as stories, much of the information is simply lost, never to be recovered. Computers, however, can keep data indefinitely. To some extent, computers can interpret data as well as store it. Some scientists think this will reduce the rate of change of human thought modes over long periods of time. It might also act to cause people in different parts of the world, and in different cultures, to think more and more alike.

Computers will make little details in information (and misinformation) more permanent. If carried to the extreme, computers will give humanity knowledge that lasts essentially forever. This has been called *immortal knowledge*. The data stored in any medium can be backed up to prevent loss because of computer failure, sabotage, and aging of disks and tapes. Every fact, every detail, and possibly all the subtle meaning, too, can be passed along unaltered for century after century.

Some engineers argue that computerization might have a detrimental effect on the preservation and accumulation of human knowledge. Computer data is easier than hard copy (such as books, scrolls, and other written documents) to tamper with on a large scale. It is not inconceivable that a few brilliant humans with nefarious intent could literally rewrite history, and no one, generations later, would be the wiser.

INCOMPLETENESS THEOREM

In 1931, a young mathematician named Kurt Gödel discovered something about logic that changed the way people think about reality. The *incompleteness theorem* demonstrated that it is impossible to prove all true statements in a first-order logical system. In any such system of thought, there are *undecidable* propositions.

In mathematical systems, certain assumptions are made. These are called *axioms* or *postulates*. Logical rules are employed to prove theorems based on the axioms. Ideally, there are no contradictions; then we have a *consistent set of axioms*. If a contradiction is found, we have an *inconsistent set of axioms*.

Generally, the stronger the set of axioms—that is, the greater the number of implied statements based on them—the greater is the chance that a contradiction can be derived. A logical system with a set of axioms that is too strong literally falls apart, because once a contradiction is found, every statement, no matter how ridiculous, becomes provable. If a set of axioms is too weak, then it does not produce much of anything meaningful. For centuries, mathematicians have striven to build "thought universes" with elegance and substance, but free of contradictions.

Gödel showed that, for any consistent set of axioms, there are more true statements than provable theorems. The set of provable statements is a proper subset of the set of all true statements, which is in turn a proper subset of the set of all possible statements (see the illustration). It follows that in any logical system without contradictions, the "whole truth" cannot be determined.

The incompleteness theorem has implications for engineers involved with *artificial intelligence (AI)*. Broadly speaking, it is impossible to build a "universal truth machine," a computer that can determine mathematically, beyond any doubt, whether any particular statement is true or false.

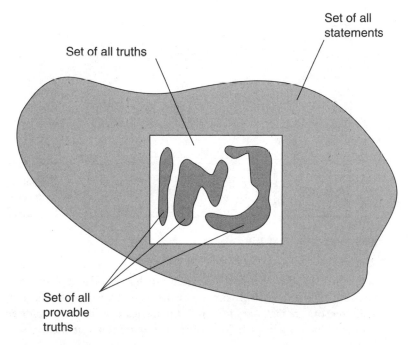

Set of all statements

Set of all truths

Set of all provable truths

Incompleteness theorem

INCREMENTAL OPTICAL ENCODER

See OPTICAL ENCODER.

INDUCTIVE PROXIMITY SENSOR

An *inductive proximity sensor* takes advantage of *electromagnetic interaction* that occurs between or among metallic objects when they are near each other.

An inductive proximity sensor uses a radio-frequency (RF) oscillator, a frequency detector, and a powdered-iron-core inductor connected into the oscillator circuit, as shown in the diagram. The oscillator is designed so a change in the magnetic flux field in the inductor core causes the frequency to change. This change is sensed by the frequency detector, which sends a signal to the apparatus that controls the robot. In this way, if the system is designed properly, a robot can avoid bumping into metallic objects. In some detectors, the flux change causes the oscillation to stop altogether. So-called metal detectors that people use to search for coins and jewelry at the beach are common examples of devices that employ inductive proximity sensors.

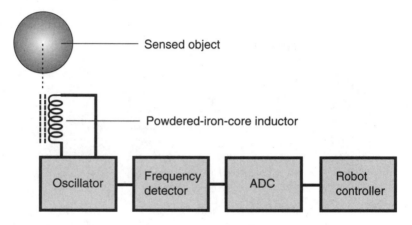

Inductive proximity sensor

Objects that do not conduct electricity, such as wood and plastic, cannot be detected by inductive proximity sensors. Therefore, other kinds of proximity sensors are necessary for a robot to navigate well in a complex environment, such as a household or office. Compare CAPACITIVE PROXIMITY SENSOR.
See also PROXIMITY SENSING.

INDUSTRIAL ROBOT

An *industrial robot*, as its name implies, is a robot employed in industry. Such robots can be fixed or mobile and can work in construction, manufacturing, packing, and quality control. They can also be used in laboratories.

Among the specific applications for industrial robots are the following: welding, soldering, drilling, cutting, forging, paint spraying, glass handling, heat treating, loading and unloading, plastic molding, bottling, canning, die casting, fruit picking, inspection, and stamping.

Two engineers, *George Devol* and *Joseph Engelberger*, were largely responsible for getting industry executives interested in robotics. Business people were hard to convince at first, but Devol and Engelberger translated things into language the business people understood: profit. The robotization of industry has not been welcomed by everyone. Humans have been displaced by robots in some industries, putting people out of work. However, the judicious use of robots in industry can improve worker safety because the machines can perform tasks that would be dangerous or deadly if done by people.

INERTIAL GUIDANCE

See GYROSCOPE.

INFERENCE ENGINE

An *inference engine* is a circuit that gives instructions to a robot. It does this by applying programmed rules to commands given by a human operator. The inference engine is something like a computer that performs *IF/THEN/ELSE* operations on a database of facts. The inference engine is the functional part of an *expert system*. See EXPERT SYSTEM and IF/THEN/ELSE.

INFINITE REGRESS

An *infinite regress* is a hypothetical scenario in which a logical process or data-transfer sequence extends backward in time indefinitely, thus having no original source. The apparent existence of an infinite regress is sometimes taken as an indication that there is something wrong with a logical argument.

Most engineers and scientists believe that computers cannot create original information. It has been assumed that meaningful data must come from outside a machine. An idea stored in a computer might come from some other computer, but if that is the case, where did the previous computer get it? From a human being, or from another computer? Based on the assumption that a computer cannot generate original thought, it follows that any idea must come either from an infinite succession of computers, one before the other before the other, without any beginning, or else from some human being. It is easier to intuit the latter scenario. Besides, computers have only been around for a few decades, so in the real world, an infinite regress of purely computer-based knowledge is an impossibility.

Some scientists have no problem with the notion that a machine can "invent" knowledge. They suggest that if a human can come up with an original thought, then a sufficiently complex machine ought to be able to do the same. Still other scientists have suggested that there is no such thing as original thought. According to this theory, even the most brilliant and insightful human beings are nothing more than sophisticated information processors, and all knowledge is the result of atomic and chemical processes in the human brain.

INFORMED COEXISTENCE

See COEXISTENCE.

INFRARED INTERFEROMETER

See PRESENCE SENSING.

INFRARED MOTION DETECTOR

See PRESENCE SENSING.

INPUT/OUTPUT MODULE

An *input/output module,* symbolized I/O, is a data link between a micro-processor and a computer's peripherals. In robotic systems, I/O modules transfer data from the controller to the mechanical parts, or vice versa. Also, I/O modules can interconnect robot controllers, or link many robots to a central computer.

The illustration shows a hypothetical example of a situation in which I/O modules are used. Triangles marked "R" are robots; circles marked

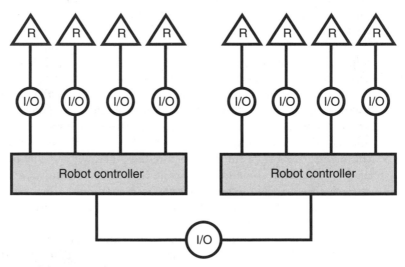

Input/output module

"I/O" are I/O modules. There are two separate robotic systems in this example, and their controllers are connected by an I/O module, allowing communication between the systems.

As its name suggests, an I/O circuit carries data in two directions: into and out of a microprocessor. It does both at the same time, so it is a *full-duplex* module.

See also CONTROLLER.

INSECT ROBOT

An *insect robot* is a member of a team of identical robots that operates under the control of a single controller, usually for the purpose of carrying out a single task or set of tasks. Such a robot is also known as a *swarm robot*. The entire group of such robots is called a *society, multiagent team,* or *swarm*. In particular, the term *insect robot* is used in reference to systems designed by engineer *Rodney Brooks*. He began developing his ideas at Massachusetts Institute of Technology (MIT) during the early 1990s.

Insect robots have six legs, and some of them actually look like beetles or cockroaches. They range in length from less than 1 mm to more than 300 mm. Most significant is the fact that they work collectively, rather than as individuals.

Autonomous robots with independent controllers are "smart in the individual," but they do not necessarily work in a team. People provide a good example. Professional sports teams have been assembled by purchasing the services of the best players in the business, but such a team rarely achieves championship status unless the players cooperate. Insects, by contrast, are "stupid in the individual." Ants and bees are like idiot robots, but an anthill or beehive is an efficient system, run by the collective mind of all its members.

Rodney Brooks saw this fundamental difference between autonomous and collective intelligence. He also saw that most of his colleagues were trying to build autonomous robots, perhaps because of the natural tendency for humans to envision robots as humanoid. To Brooks, it was obvious that a major avenue of technology was being neglected. Thus he began designing robot teams, consisting of many units with a single controller.

Brooks is a futurist who envisions microscopic insect robots that might live in your house, coming out at night to clean your floors and countertops. "Antibody robots" of even tinier proportions could be injected into a person infected with some heretofore-incurable disease. Controlled by a central microprocessor, they could seek out the disease bacteria or viruses and swallow them up. Compare AUTONOMOUS ROBOT.

INTEGRAL

The term *integral* refers to the area under the curve of a mathematical function. For example, displacement is the integral of speed or velocity, which in turn is the integral of acceleration.

Figure 1 shows a generalized graph of speed as a function of time. The curve of this function is a straight line. It might represent the steady increase in speed of an accelerating mobile robot. While the speed constantly increases, the displacement, indicated by the area under the curve, increases at a faster rate.

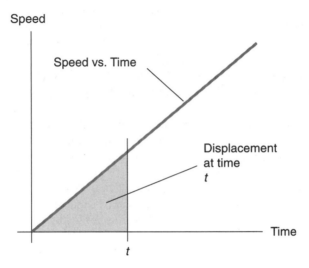

Integral—Fig. 1

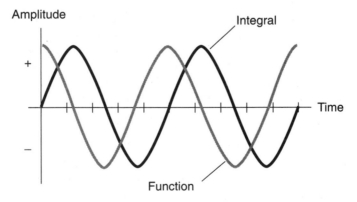

Integral—Fig. 2

In digital electronics, a circuit that continuously takes the integral of an input wave is called an *integrator*. An example of the operation of an integrator is shown in the graph of Fig. 2. The input is a sine wave. The output is, mathematically, a negative-cosine wave, but it appears as a sine wave that has been shifted by 90°, or one-fourth of a cycle. Compare DERIVATIVE.

INTEGRATED CIRCUIT

An *integrated circuit* (IC) is an electronic device containing many diodes, transistors, resistors, and/or capacitors fabricated onto a wafer, or chip, of semiconductor material. The chip is enclosed in a small package with pins for connection to external components. Integrated circuits are used extensively in robots and their controllers.

Assets and limitations

Integrated-circuit devices and systems are far more compact than equivalent circuits made from discrete components. More complex circuits can be built, and kept down to a reasonable size, using ICs as compared with discrete components. Thus, for example, there are notebook computers with capabilities more advanced than early computers, which took up entire rooms.

In an IC, the interconnections among components are physically tiny, making high switching speeds possible. Electric currents travel fast, but not instantaneously. The faster the charge carriers move from one component to another, the more operations can be performed per unit time, and the less time is required for complex operations.

Integrated circuits consume less power than equivalent discrete-component circuits. This is important if batteries are used. Because ICs draw so little current, they produce less heat than their discrete-component equivalents. This results in better efficiency, and minimizes problems that plague equipment that gets hot with use, such as frequency drift and generation of internal noise.

Systems using ICs fail less often, per component-hour of use, than systems that make use of discrete components. This is mainly because all interconnections are sealed within an IC case, preventing corrosion or the intrusion of dust. The reduced failure rate translates into less downtime.

Integrated-circuit technology lowers service costs, because repair procedures are simple when failures occur. Many systems use sockets for ICs, and replacement is simply a matter of finding the faulty IC, unplugging it, and plugging in a new one. Special desoldering equipment is used for servicing circuit boards that have ICs soldered directly to the foil.

Modern IC appliances employ *modular construction*. Individual ICs perform defined functions within a circuit board; the circuit board or card, in turn, fits into a socket and has a specific purpose. Computers, programmed with customized software, are used by technicians to locate the faulty card in a system. The card can be pulled and replaced, getting the system back to the user in the shortest possible time.

Linear ICs

A *linear integrated circuit* is used to process analog signals such as voices, music, and most radio transmissions. The term "linear" arises from the fact that the instantaneous output is a linear function of the instantaneous input.

An *operational amplifier* (also called an *op amp*) consists of several transistors, resistors, diodes, and capacitors, interconnected to produce high gain over a wide range of frequencies. An op amp has two inputs and one output. When a signal is applied to the *noninverting input,* the output is in phase with it; when a signal is applied to the *inverting input,* the output is 180° out of phase with it. An op amp has two power supply connections, one for the emitters of the transistors (V_{ee}) and one for the collectors (V_{cc}). The symbol for an op amp is a triangle. The inputs, output, and power supply connections are drawn as lines emerging from the triangle. The gain characteristics of an op amp are determined by external resistors. Normally, a resistor is connected between the output and the inverting input. This is the *closed-loop configuration.* The feedback is negative, causing the gain to be less than it would be if there were no feedback (*open-loop configuration*).

A closed-loop amplifier using an op amp is shown in Fig. 1. When a resistor–capacitor (*RC*) combination is used in the feedback loop of an op amp, the amplification factor varies with the frequency. It is possible to get a low-pass response, a high-pass response, a resonant peak, or a resonant notch using an op amp and various *RC* feedback arrangements.

A *voltage-regulator* IC acts to control the output voltage of a power supply. This is important with precision electronic equipment. These ICs are available in various different voltage and current ratings. Typical voltage-regulator ICs have three terminals. They look like power transistors.

A *timer* IC is a form of oscillator. It produces a delayed output, with the delay being variable to suit the needs of a particular device. The delay is generated by counting the number of oscillator pulses. The length of the delay is adjusted by means of external resistors and capacitors.

A *multiplexer* IC allows several different signals to be combined in a single channel via time-division multiplexing, in a manner similar to that

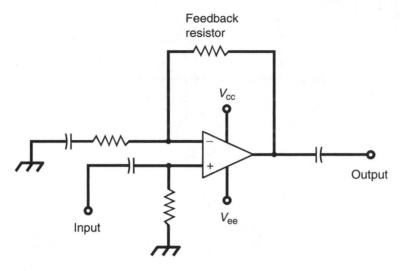

Integrated circuit—Fig 1

used with pulse modulation. An analog multiplexer can also be used in reverse; then it works as a *demultiplexer*.

Like an op amp, a *comparator* IC has two inputs. The device compares the voltages at the two inputs (called A and B). If the input at A is significantly greater than the input at B, the output is about +5 V. This is logic 1, or high. If the input at A is not greater than the input at B, the output voltage is about +2 V. This is designated as logic 0, or low. Comparators are employed to actuate, or trigger, other devices such as relays and electronic switching circuits. They have various applications in robotic systems.

Digital ICs

Digital integrated circuits consist of gates that perform logical operations at high speeds. There are several different technologies, each with unique characteristics. Digital-logic technology can employ bipolar and/or metal-oxide semiconductor devices.

In *transistor-transistor logic* (TTL), arrays of bipolar transistors, some with multiple emitters, operate on DC pulses. A TTL gate is illustrated in Fig. 2. The transistors are either cut off or saturated; there is no "in between." Because of this, TTL circuitry is comparatively immune to extraneous noise.

Another bipolar-transistor logic form is known as *emitter-coupled logic* (ECL). In ECL, the transistors are not operated at saturation, as they are with TTL. This increases the speed of operation of ECL compared with TTL. However, noise pulses have a greater effect in ECL, because

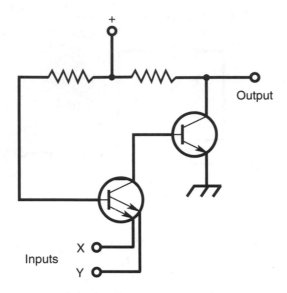

+

Output

X

Inputs

Y

Integrated circuit—Fig 2

unsaturated transistors amplify as well as switch signals. The schematic of Fig. 3 shows a simple ECL gate.

N-channel metal-oxide-semiconductor (NMOS) logic offers simplicity of design, along with high operating speed. *P-channel metal-oxide-semiconductor* (PMOS) logic is similar to NMOS, but the speed is slower. An NMOS or PMOS digital IC is like a circuit that uses only N-channel field-effect transistors (FETs), or only P-channel FETs.

Complementary-metal-oxide-semiconductor (CMOS) logic employs both N-type and P-type silicon on a single chip. This is analogous to using N-channel and P-channel FETs in a circuit. The main advantages of CMOS are extremely low current drain, high operating speed, and immunity to noise.

INTEGRATION

See INTEGRAL.

INTELLIGENT COEXISTENCE

See COEXISTENCE.

INTEREST OPERATOR

In machine vision, an *interest operator* is an algorithm that selects "interesting" pixels (picture elements) in the image. "Interesting" in this context refers to pixels that are different from the majority of those in their vicinity.

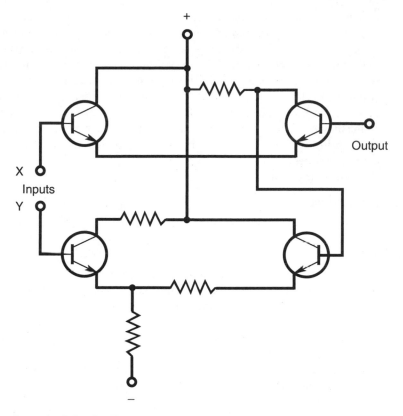

Integrated circuit—Fig 3

Examples include dark or bright spots, different colors, sharp edges, lines, and curves.

Interest operators can serve various purposes in a robotic vision system. One function is to eliminate *correspondence,* an undesirable condition that can occur in *binocular machine vision* when patterns confuse the machine's sense of depth. Interest operators provide points of reference independent of general patterns. Another function of an interest operator is to ascertain boundaries, as in *edge detection.* Such algorithms are also useful in vision systems that incorporate *local feature focus.*

See also BINOCULAR MACHINE VISION, CORRESPONDENCE, LOCAL FEATURE FOCUS, and VISION SYSTEM.

INTERFACE

An *interface* is a device that carries data between a computer and its peripherals, or between a computer and a human. An interface consists of both hardware and software. The term is also used as a verb; when you connect two devices together and make them compatible, you interface them.

Suppose you want to use a computer to control a robot. You must ensure that they will work together. That is, you must interface the computer with the robot. This requires the correct medium for transferring the data (cable or wireless link), the use of the proper type of data port (parallel or serial), and the correct program for robot control. In a robotic system, all the moving parts are, in effect, peripherals to the controller, in the same way as printers, scanners, and external drives are peripherals in a personal computer system.

See also CONTROLLER.

INTERFEROMETER

See PRESENCE SENSING.

INTERPOLATION

When there is a gap in data, but data are available on either side of the gap, an estimate of values within the gap can sometimes be made by means of a mathematical process called *interpolation.*

Figure 1 shows a smooth curve with a gap in the plotted, or known, values. The simplest way to interpolate the values within the unknown region is to connect the "loose ends" of the curve with a straight line (shown as a dashed line in this example). This is called *linear interpolation.* More sophisticated schemes attempt to derive a function that defines the

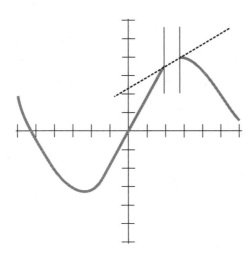

Interpolation—Fig. 1

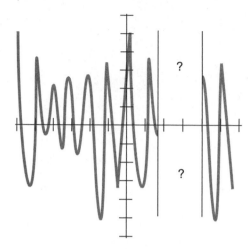

Interpolation—Fig. 2

curve in the vicinity of the gap, and then fill in the unknown values based on this function.

Simple interpolation methods do not necessarily work for complex waveforms, especially when the gap is large (Fig. 2). Unless a pattern can be determined for the curve, thereby making it possible to derive a mathematical function that represents the curve, precise points within the gap cannot be determined. Even if it appears that a function has been found to define a complex or irregular curve, interpolation may not work because the derived function is based on an insufficient sampling of the data. Compare EXTRAPOLATION.

INTRUSION DETECTION

See PRESENCE SENSING.

J

JAW

A *jaw* is a specialized robot gripper, consisting of individual parts that can clamp down to hold an object, and open to let the object go free. The device gets its name from its functional resemblance to the human jaw, or to the jaws of various animals and insects.

A typical robotic jaw has two parts that are hinged at a common end. One or both parts can move relative to the robot arm. Some jaws have three or four movable parts that come together to grip an object, and dilate to free it.

See also ROBOT GRIPPER.

JOIN

The term *join* is used in reference to a controller programming function that allows a robot to resume a task where it left off, in the event of a disruption such as a power failure or accident. This feature is similar to, although generally more sophisticated than, the capability of a computer printer to start printing at the page where it left off, if the paper tray becomes empty during a job and must be refilled.

An effective join program requires a *nonvolatile memory,* such as random-access memory (RAM) with battery backup, to store information concerning actions in the task that have already been performed, and those that have yet to be performed. The present moment, or instant in time, must be clearly known by the robot controller and updated at frequent intervals (for example, a fraction of a second). This information is constantly stored and refreshed in the nonvolatile RAM.

If there is a power interruption, accident, or other mishap, the robot is programmed to go through a certain sequence to determine where and how to begin again, based on the data stored in RAM. Along with the sequence of previously executed movements and the movements yet to be performed in a given task, additional information might be required, such as whether the physical location or orientation of the robot has changed within the work environment.

JOINTED GEOMETRY

See ARTICULATED GEOMETRY.

JOINT-FORCE SENSING

Joint-force sensing keeps a robot joint from exerting too much force. A feedback system is used. The sensor works by detecting the resistance the robot arm encounters. As the applied force increases, so does the resistance. The sensor is programmed to reduce or stop the joint if a set amount of resistance is exceeded.

See also BACK PRESSURE SENSOR.

JOINT-INTERPOLATED MOTION

In a robot arm having more than one joint, the most efficient mode of operation is known as *joint-interpolated motion*. In this scheme, the joints move in such a way that the *end effector* reaches the required point at the exact instant that each of the joints has completed its assigned motion.

In order for a multijointed robot arm to position the end effector at a designated location, each joint must turn through a certain angle. (For some joints this angle might be zero, representing no rotation.) The designated location can be reached by any sequence of events such that each joint rotates through its assigned angle; the same end point will result whether or not the joints move at the same time. For example, each joint can rotate through its assigned angle while all the others remain fixed, but this is a time-consuming and inefficient process. The fastest and most efficient results are obtained when all the joints begin rotating at a certain instant in time t_0, and all of them stop rotating at a certain instant t_1, which is $(t_1 - t_0)$ later than t_0.

Suppose a robot arm using *articulated geometry* has three joints that rotate through angles $X = 39$ degrees, $Y = 75$ degrees, and $Z = 51$ degrees, as shown in the illustration. Suppose the end effector is programmed to reach its end point exactly 3 seconds after the joints begin to rotate. If the joints rotate at the angular speeds shown (13, 15, and 17 degrees per second, respectively), the end effector arrives at its designated stopping point precisely when each joint has turned through its required angle. This is an example of joint-interpolated motion.

See also ARTICULATED GEOMETRY and DEGREES OF ROTATION.

JOINT PARAMETERS

The *joint parameters* of a robot arm or *end effector* are the scalar values, usually measured in linear displacement units and angular units, all of

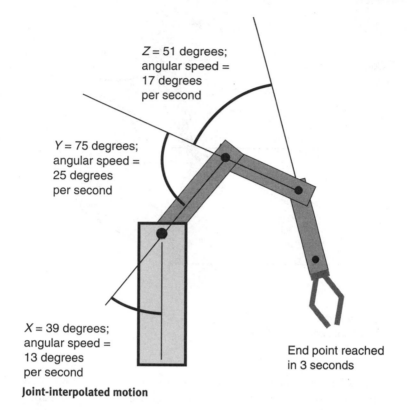

Z = 51 degrees;
angular speed =
17 degrees
per second

Y = 75 degrees;
angular speed =
25 degrees
per second

X = 39 degrees;
angular speed =
13 degrees
per second

End point reached
in 3 seconds

Joint-interpolated motion

which together define the set of all possible positions the device can attain.

As an example, suppose a robot arm has three joints, each of which can rotate over 180°, along with a swivel base that can rotate 360°. This robot arm has four joint parameters, assuming the swivel base is considered as a joint.

See also ARTICULATED GEOMETRY, DEGREES OF FREEDOM, and DEGREES OF ROTATION.

JOYSTICK

A *joystick* is a control device capable of movement in two or three dimensions. The device consists of a movable lever or handle, and a ball-bearing within a control box. The stick is moved by hand.

The joystick gets its name from its resemblance to the joystick in an aircraft. Some joysticks can be rotated clockwise and counterclockwise, in addition to the usual two coordinates, allowing control in three dimensions, labeled x, y, and z. The illustration shows an example of such a device. A button switch can be placed at the top end of the movable lever, allowing limited control in a fourth dimension (w).

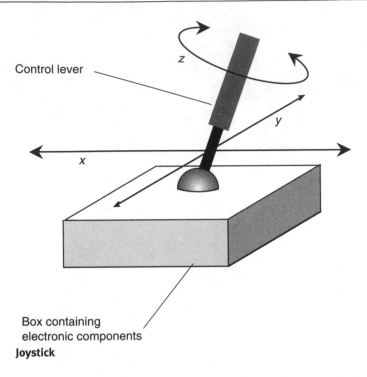

Control lever

z

y

x

Box containing
electronic components
Joystick

Joysticks are used in computer games, for entering coordinates into a computer, and for the remote control of robots. Many of these are more sophisticated than the basic device shown here; some require gripping with one or both hands.

JUNGIAN WORLD THEORY

An interesting motivation for *artificial intelligence (AI)* research is called *Jungian world theory.* According to this hypothesis, human beings keep making the same mistakes in every generation. It seems that people cannot learn from history. Humanity, as a collective unit, seems unable to foretell, or care about, the potential consequences of the things they do. It is as if the human race is time-blind. Thus, "History repeats."

This theory has been demonstrated many times. People keep fighting wars for the same reasons. Wars rarely solve problems, although in some cases it seems as if there is no other choice than to go to war. Human activities become ever more destructive to Earth's ecosystem. The most pessimistic interpretations of Jungian world theory suggest that humanity is doomed to self-extinction.

What can humanity do to stop this self-defeating vicious circle? According to researcher *Charles Lecht,* one answer may lie in the development of

AI to the point that machines attain greater intelligence than people. Perhaps a brilliant computer or system of machines can help humanity to control its destiny, so that people need not keep reliving the same old calamities.

Many researchers doubt that machines will, or can, become smarter than people, but it has been argued that AI can and should be used to help humanity find solutions to difficult social problems.

K

KINEMATIC ERROR

Kinematic error refers to imprecision in robot motion that takes place independently of force and mass. The ultimate effect of kinematic error can be measured in absolute terms, such as linear displacement units or degrees of arc. The ultimate effect can also be measured in terms of a percentage of the total movement.

As an example, suppose a mobile robot is programmed to proceed at a speed of 1.500 m/s at an azimuth bearing of 90.00° (due east) on a level surface. If the robot encounters an upward incline, the forward speed can be expected to decrease slightly. If the robot encounters a downslope, in contrast, the forward speed can be expected to increase. If the surface banks to the left or the right, the direction of motion can be expected to change accordingly. In the ideal scenario, terrain irregularity would not affect the speed or direction of the machine; the kinematic error would therefore be zero.

Kinematic errors, if they take place persistently in a given sense and over a period of time, can result in accumulation of the displacement or position of the robot when the task is completed. Compare DISPLACEMENT ERROR.

K-LINE PROGRAMMING

K-line programming is a method by which a smart robot can learn as it does a job, so that it will have an easier time doing the same or similar work in the future.

Suppose you have a *personal robot* that you use for handiwork around the house. The water heater breaks down and you instruct the robot to fix it. The robot must use certain tools to do the repair. The first time the robot repairs the water heater, it must find the tools by trial and error. It encodes each tool in its memory, perhaps according to shape. It also encodes the sequence in which the tools are used to fix the water heater. The list of tools used before, and the order in which they were used, is called a K line. The next time the water heater needs repair, the robot can refer to the K line to streamline the process of executing the task.

Of course, there are many different things that can go wrong with a water heater. The second time the water heater breaks down, the K line for the first repair might not work. In that sort of case, the robot must refine its knowledge, devising a second K line for the new problem. Over time, the robot will learn several different schemes for fixing a water heater, each scheme tailored to a specific problem. The illustration is a flowchart showing how a repertoire of K lines can be developed so the robot learns by experience.

See also HEURISTIC KNOWLEDGE.

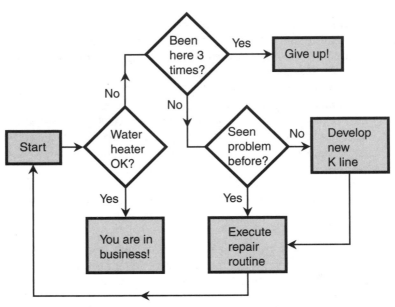

K-line programming

KLUDGE

A crude, useless, or grossly inefficient device or process is called a *kludge* (pronounced "kloodge"). The term is also used in reference to a temporary fix or patch. You might say, "That contraption is a kludge robot," or "This is a kludge that will make the program run more smoothly."

Kludges are often useful, because they can test an idea without a lot of trouble and expense. But sometimes, referring to a device or scheme as a "kludge" is an apology (in the case of one's own work) or a good-natured insult (in the case of someone else's work).

Most experienced engineers have occasionally built or written kludge devices or programs. Every year, Purdue University holds a "Rube Goldberg" contest for students to build or design ridiculously inefficient machines and computer software. The contest is sponsored by *Theta Tau,* an engineering fraternity.

Eventually, almost every engineer becomes an expert at the art of kludge. But trying to sell a kludge as a finished product is an industrial error—unless, of course, the intent is to make a joke.

KNOWLEDGE

The term *knowledge* refers to the data stored in a computer system, robot controller, or human mind. Also, the term refers to how well a brain, either electronic or biological, makes use of the data it has.

Humans, individually and collectively, have knowledge that changes from generation to generation. Some researchers have suggested that computers, along with electronic, optical, and magnetic storage media, will eliminate the loss or degradation of human knowledge in future generations. This will give humanity an ever-expanding storehouse of *immortal knowledge.*

In *expert systems,* computer engineers define *knowledge acquisition* as the process by which machines obtain data. Generally, it is agreed that all computer knowledge must come from human beings, although a few scientists believe that machines can generate original, true knowledge. Although there is controversy about the ability of machines to create original knowledge, it has been conclusively demonstrated that high-level computers can learn from their mistakes. This is not original thought, but is derived from existing knowledge by programming. The ability of a machine to improve the use of its data is called *heuristic knowledge.*

Computers can store and manipulate information in ways that people find difficult or impossible. A good example is the addition of a series of 5 million numbers. However, there are problems humans can solve that a machine cannot, and perhaps will never, be able to figure out. One example of this is the regulation of the amount of medication needed to keep a hospital patient anesthetized during surgery, without causing harm to the patient.

See also EXPERT SYSTEM, HEURISTIC KNOWLEDGE, IMMORTAL KNOWLEDGE, and INFINITE REGRESS.

L

LADAR

Ladar is an acronym that stands for *laser detection and ranging*. It is also known as *laser radar* or *lidar* (short for *light detection and ranging*).

In robotics, a ladar system uses a shaft of visible light or infrared (IR) energy, rather than radio waves (as in *radar*) or acoustic waves (as in *sonar*) to perform *range sensing and plotting* of the environment. The device works by measuring the time it takes for a laser beam to travel to a target point, reflect from it, and then propagate back to the point of transmission.

The principal asset of ladar over other ranging methods is the fact that the laser beam is extremely narrow. This provides vastly superior *direction resolution* compared with radar and sonar schemes, whose beams cannot be focused with such precision.

Ladar has limitations. It cannot work well through fog, or precipitation, as can radar. Certain types of objects, such as mirrors oriented at a slant, do not return ladar energy and produce no echoes.

A high-level ladar system scans both horizontally and vertically, thereby creating a three-dimensional (3-D) *computer map* of the environment. This type of system is extremely expensive. Less sophisticated ladar devices work in a single plane, usually horizontal, to create a two-dimensional (2-D) computer map of the environment at a specific level above a floor or flat ground. Compare RADAR and SONAR.

See also COMPUTER MAP, DIRECTION RESOLUTION, and RANGE SENSING AND PLOTTING.

LADLE GRIPPER

A *ladle gripper* is a robotic *end effector* that can be used to move liquids. It can also be used to move powders and gravel. The device gets its name from its scooplike physical shape, and the manner in which it operates. The end effector can be shaped like a half-sphere, a box, or any other container that can hold liquids. Ladle grippers are used in industry to move molten metals from vats into molds. They can also be used to handle certain types of hazardous material.

Ladle grippers require significant gravitation or acceleration force in which to operate. For this reason, they are generally not suitable for use in outer space. Also, the material being moved must have a significantly greater density than the medium in which the movement takes place. For example, a ladle gripper might be used to move ethanol from one container to another in air at sea level, but not under water.

See also ROBOT GRIPPER.

LANDMARK

In robotics, a *landmark* is a specific feature of a robot's work environment, notable because of its usefulness in navigation and ranging. Landmarks are generally fixed with respect to time. Examples include a desk, a doorway, or a set of objects such as buildings or signs. Landmarks can be natural or artificial. Sometimes they are deliberately positioned for the purpose of assisting robots in their navigation within a region.

An imaginary line between two landmarks in called a *landmark pair boundary.* Unless there are obstructions or hazards, landmark pair boundaries are usually straight. In a complex work environment, landmark pair boundaries form the edges of triangles known as *orientation regions.* An example is shown in the illustration.

See also COMPUTER MAP, RELATIONAL GRAPH, and TOPOLOGICAL PATH PLANNING.

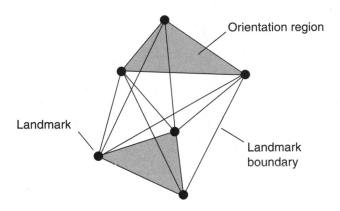

Landmark

LASER DATA TRANSMISSION

Laser beams can be modulated to convey information, in the same way as radio waves. *Laser data transmission* allows a few broadband signals, or many narrowband signals, to be sent over a single beam of light. This method of data transmission is used in some mobile robotic systems.

A laser-communications transmitter has a signal processor or amplifier, a modulator, and laser (upper part of illustration). The receiver uses a photocell, an amplifier, and a signal processor (lower part of illustration). Any form of data can be sent, including voice, television, and digital signals. Laser communications systems can be either line-of-sight or fiber-optic.

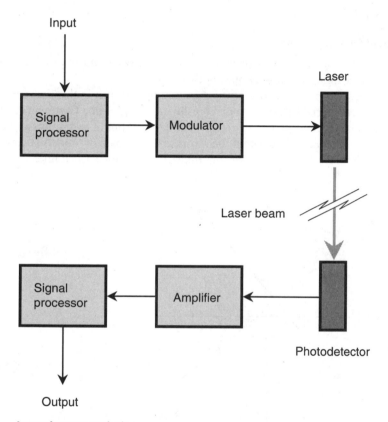

Laser data transmission

In a line-of-sight system, the beam travels in a straight line through space or clear air. Because laser beams remain narrow for long distances, long-range communication is possible. However, this scheme does not work well through clouds, fog, rain, snow, or other obstructions. The alignment of the laser and the photodetector must be precise.

In a fiber-optic laser-communications system, the beam is guided through a glass or plastic filament. This is similar to wire or cable communications, but with far more versatility. Optical systems are not susceptible to the effects of electromagnetic interference (EMI), as are some wire and cable networks. Fiber-optic systems lend themselves well to robot control, especially in hostile environments such as the deep sea. Compare MICROWAVE DATA TRANSMISSION.

LEGGED LOCOMOTION

See ROBOT LEG.

LINEAR PROGRAMMING

Linear programming is a process of optimizing two or more variables that change independently of one another.

A simple example of linear programming is shown in the illustration. The two variables are position coordinates, x and y, in a Cartesian (rectangular) plane. The variables represent the positions of two robots as they move in straight lines within their work area. The path of robot A is shown by the solid line; the path of robot B is shown by the dashed line.

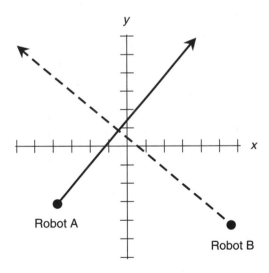

Linear programming

Suppose robot A moves at 1.150 m/s and robot B moves at 0.755 m/s. The starting points are shown by heavy dots. Linear programming can answer these questions:

- How long after their start-up will the robots be closest to each other?
- Will the robots collide unless they change course or speed?
- What will be the coordinates of robot A at the time of closest approach?
- What will be the coordinates of robot B at the time of closest approach?

Computers can be programmed to solve these problems quickly and easily. Smart robots solve such problems when necessary, without oversight by a human operator. These calculations are important if there are numerous robots in a small work area and the robots lack sophisticated collision-avoidance sensors.

LOAD/HAUL/DUMP

Load/haul/dump, abbreviated LHD, is a type of mobile robot used in mining and construction. It does exactly what its name implies. With the aid of a human operator, LHDs load cargo, haul it from one place to another, and dump it in a prescribed location.

In mining, LHDs have an easier time than in general construction. The geometry of a mine is easily programmed into the robot controller; the layout changes slowly. Reprogramming does not need to be done often. In construction, however, the landscape is more complicated, and it changes rapidly as work progresses. Therefore, the computer maps must be revised often. In mining, all the loads are usually the same in terms of weight and volume, because each load consists of a prescribed amount of a single substance such as coal or iron ore. In construction, the nature, and thus the weight and volume, of the load can vary.

LHD vehicles use various methods for navigation, including beacons, computer maps, position sensors, and vision systems. LHDs can be autonomous, although there are advantages to using a single controller for many robots.

See also AUTONOMOUS ROBOT and INSECT ROBOT.

LOCAL FEATURE FOCUS

In a robotic vision system, it is usually not necessary to use the whole image to perform a function. Often, only a single feature, or a small region within the image, is needed. To minimize memory space and to optimize speed, *local feature focus* can be used.

Suppose a robot needs to get a pliers from a toolbox. This tool has a characteristic shape that is stored in memory. Several different images can be stored, representing the pliers as seen from various angles. The vision system quickly scans the toolbox until it finds an image that matches one of its images for pliers. This saves time compared with trial and error, in which the robot picks up tool after tool until it finds pliers.

The human eye/brain system uses local focus without conscious effort. If someone is driving in a forest and sees a sign that says, "Watch for animals crossing the road," the driver will be on the lookout for animals on or near the roadway. A tractor parked on the shoulder will not arouse interest, but a horse will. The human eye/brain system can instantly tell animate from inanimate objects. A robot controller with sophisticated local feature focus, in conjunction with *artificial intelligence (AI)*, can do the same.

See also OBJECT RECOGNITION and VISION SYSTEM.

LOGIC

Logic can refer to either of two things in electronics, computer science, and *artificial intelligence (AI)*.

Boolean algebra is the representation of statements as symbols, along with operations, generating equations. This form of logic is important in the design of digital circuits, including computers. Boolean algebra is a form of *deductive logic,* because the conclusions are derived, or deduced, in a finite number of steps.

In *mathematical induction,* a statement is proven true for a sequence of cases. First, the statement is proved deductively for one case. Then it is proved that if the statement is true for some arbitrary case, it is true for the next case in the sequence. This implies truth for the whole sequence, even if the sequence is infinite. Mathematicians consider this perfectly rigorous and acceptable. For a thorough discussion of deductive logic and mathematical induction, a text on symbolic logic is recommended.

Trinary logic allows for a neutral condition, neither true nor false, in addition to the usual true/false (high/low) states. These three values are represented by logic -1 (false), 0 (neutral), and $+1$ (true). Trinary logic can be easily represented in electronic circuits by positive, zero, and negative currents or voltages.

In *fuzzy logic,* values cover a continuous range from "totally false," through neutral, to "totally true." Fuzzy logic is well suited for the control of certain processes. Its use will become more widespread as AI technology advances. Fuzzy logic can be represented digitally in discrete steps; the number of steps is usually some power of 2.

See also BOOLEAN ALGEBRA and LOGIC GATE.

LOGIC GATE

All binary digital devices and systems employ electronic switches that perform various Boolean functions. A switch of this type is called a *logic gate*.

Usually, the binary digit 1 stands for "true" and is represented by about +5 V. The binary digit 0 stands for "false" and is represented by a voltage near 0 V. This scheme is known as *positive logic*. There are other logic forms, the most common of which is *negative logic* (in which the digit 1 is represented by a more negative voltage than the digit 0). The remainder of this discussion deals with positive logic.

An *inverter*, also called a *NOT gate*, has one input and one output. It reverses the state of the input. An *OR gate* can have two or more inputs. If both, or all, of the inputs are 0, then the output is 0. If any of the inputs are 1, then the output is 1. An *AND gate* can have two or more inputs. If both, or all, of the inputs are 1, then the output is 1. Otherwise the output is 0.

Sometimes an inverter and an OR gate are combined. This produces a *NOR gate*. If an inverter and an AND gate are combined, the result is a *NAND gate*. An *exclusive OR gate*, also called an *XOR gate*, has two inputs and one output. If the two inputs are the same (either both 1 or both 0), then the output is 0. If the two inputs are different, then the output is 1.

The functions of logic gates are summarized in the accompanying table, and their schematic symbols are shown in the illustration.

See also BOOLEAN ALGEBRA and LOGIC.

Logic gate: common types and their characteristics

Gate type	Number of inputs	Remarks
INVERTER (NOT)	1	Changes state of input
OR	2 or more	Output high if any inputs are high
		Output low if all inputs are low
AND	2 or more	Output low if any inputs are low
		Output high if all inputs are high
NOR	2 or more	Output low if any inputs are high
		Output high if all inputs are low
NAND	2 or more	Output high if any inputs are low
		Output low if all inputs are high
XOR	2	Output high if inputs differ
		Output low if inputs are the same

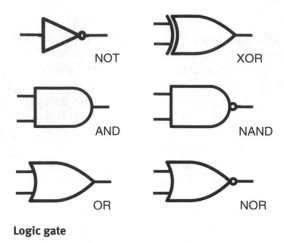

Logic gate

LOG-POLAR TRANSFORM

When mapping an image for use in a robotic vision system, it sometimes helps to transform the image from one type of coordinate system to another. In a *log-polar transform,* a computer converts an image in polar coordinates to an image in rectangular coordinates.

The principle of log-polar image processing is shown in the illustration. The polar system, with two object paths plotted, is depicted in the upper graph. The rectangular equivalent, with the same paths shown, is in the lower graph. The polar radius is mapped onto the vertical rectangular axis; the polar angle is mapped onto the horizontal rectangular axis.

Radial coordinates are unevenly spaced in the polar map, but are uniform in the rectangular map. During the transformation, the logarithm of the radius is taken. This results in peripheral distortion of the image. The resolution is degraded for distant objects, but is improved for nearby targets. In robotic navigation, close-in objects are usually more important than distant ones, so this is a good trade-off.

A log-polar transform greatly distorts the way a scene looks to people. A computer does not have any trouble with this, because each point in the image corresponds to a unique point in real space. That is, the point mapping is a *one-to-one correspondence.*

Robot vision systems employ television cameras that scan in rectangular coordinates, but events in real space are of a nature better represented by polar coordinates. The log-polar transform can therefore change real-life motions and perceptions into images that can be efficiently dealt with by a robotic vision system.

See also VISION SYSTEM.

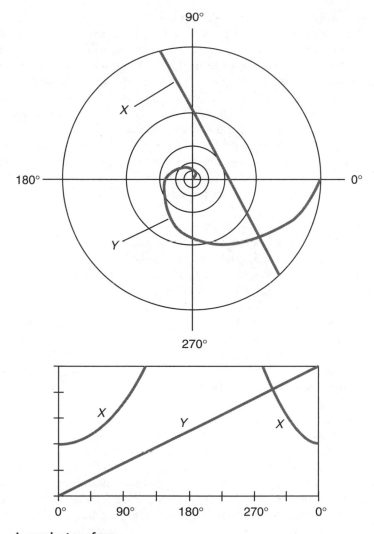

Log-polar transform

LOOP

A *loop* is a repeating sequence of operations in a computer program. The number of repetitions can range from two to thousands, millions, or billions. Often, the number of iterations depends on the data input. In some programs, there are loops within loops. This is called *nesting of loops*.

Loops are useful in mathematical calculations that involve repetitious operations. Until computers were developed, many such problems could not be solved. The problems did not involve esoteric principles, but the trillions upon trillions of steps would take a single person, even equipped with a powerful calculator (or an abacus), more than a lifetime to grind out.

Sometimes errors are made in the programming, and a computer ends up going through a loop without ever reaching a condition in which it can exit the loop. This is called an *infinite loop* or *endless loop*. It always results in a failure of the program to come to a satisfactory conclusion. In the extreme, it can cause the computer to crash.

In any system involving feedback, the path of the feedback signal is called the *loop*. The term *loop* can also refer to a graphical rendition of the operation of a system that employs feedback.

See also FEEDBACK and SERVOMECHANISM.

LUDDITE

Whenever there is a major new technological innovation, some people fear that they will lose their jobs. Job loss can occur for at least two reasons. First, greater efficiency reduces the number of people needed for a corporation or agency to function. Second, human workers have occasionally been replaced by machines because they do not get sick, do not take coffee breaks, and do not demand vacation time. People who have an exaggerated fear of technology for any reason are called *technophobes*.

During the Industrial Revolution in England, technophobes went on rampages and destroyed new equipment that they feared would take their jobs from them. Their leader was a man named *Ned Ludd,* and so these people became known as *Luddites.*

Robotization has not caused a latter-day Luddite-type reaction in the United States, Japan, or Europe. The reasons for this are not completely known. Some roboticists suggest that the absence of a major Luddite movement today is due to the fact that the standard of living is higher now than it was in Ned Ludd's time. Society is, by all indications, dependent on computers, robots, and other high-tech devices, and everyone—even the technophobes—know that destroying these machines would do more harm than good.

M

MACHINE LANGUAGE

A computer does not work with words, or even with familiar base-10 numbers. Instead, the machine uses combinations of ones and zeros. These are the two binary states, also represented by on/off, high/low, or true/false. Data in *machine language*, if written down, looks like a string of ones and zeros, such as 0110100100. This can be represented pictorially by a graph as shown in the illustration.

When a computer operator writes a program, or a person issues a command to a robot controller, it is done in a *high-level language*. This must be converted into machine language for the computer. The computer output is likewise translated from machine language into whatever high-level language is used by the programmer or operator.

See also HIGH-LEVEL LANGUAGE.

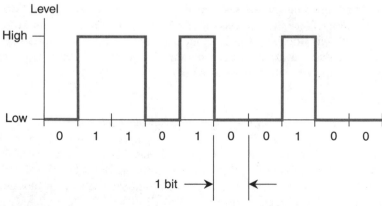

Machine language

MACHINING

In industrial robotics, *machining* is the modification of parts during assembly. Examples of machining are drilling, removing burrs from drilled holes, welding, sanding, and polishing. In an assembly line, many identical parts pass each workstation in rapid succession, and the worker or robot does the same tasks repeatedly.

There are two methods by which machining is done by robots. The robot can hold the tool while the part remains stationary, or the robot can hold the part while the tool stays put.

Robot holds tool

This method offers the following advantages:

- Small robots can be used if the tool is not heavy.
- Parts can be large and heavy, because they need not be moved by the robot.
- The robot can adjust easily as the tool wears down.

Robot holds part

In this method, the advantages are:

- The part can be moved to any of several different tools, without having to change the tool on the robot arm.
- Tools can be large and heavy, because they need not be moved.
- Tools can have massive, powerful motors because the robot does not have to hold them.

Some industrial situations lend themselves better to the first method, while some processes are done more efficiently using the second method.

There are some processes that do not lend themselves to robotic machining. These include tasks that require subjective decisions. Some products will probably never be made using robots, because it will not be cost-effective. An example is a custom-built automobile, put together part by part rather than on an assembly line.

MACROKNOWLEDGE

Macroknowledge is a term used in *artificial intelligence (AI)* that means "knowledge in the large sense." An example of macroknowledge is a set of definitions for different classes of living things. The two main classes are plants and animals (although some life forms share characteristics of both classes). Within the class of animals, we might focus on warm-blooded versus cold-blooded creatures.

Macroknowledge about living things might be used by a smart robot to determine, for example, whether a biped approaching it is a human, another robot, or a gorilla. Compare MICROKNOWLEDGE.

MAGNITUDE PROFILE

The term *magnitude profile* refers to the way in which a robot behaves near an object of interest. In particular, the term refers to the variation in the length (magnitude) of a vector, depending on the distance (radius) from the object of interest. The vector magnitude might represent the output level from a *proximity sensor* or *distance-measurement* device, or the robot's speed or acceleration in a particular direction relative to the object of interest.

As an example, suppose a robot is equipped with a proximity sensor designed to warn it when obstructions are nearby. The output from the sensor increases as the distance between the robot and the object decreases. This can take place according to various magnitude profiles. The illustration graphs three of the most common.

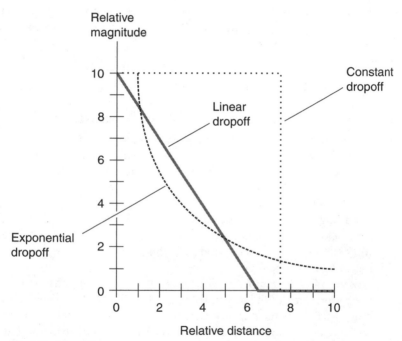

Magnitude profile

In a system that exhibits a *constant-dropoff profile*, which can also be called the *threshold-detection profile*, the sensor output is zero until the robot comes within a specific distance of the object (in this case approximately 7.5 units as shown on the graph). When the robot is closer than the critical radius, the sensor output is high and constant, and does not vary with the distance.

In a system with a *linear-dropoff profile*, the sensor output is zero until the robot comes within a specific distance of the object (in this case approximately 6.5 units as shown on the graph). When the robot is closer than the critical radius, the sensor output varies according to a straight-line function with negative slope, as shown, reaching a maximum when the robot is about to strike the object.

In a system with an *exponential-dropoff profile*, the sensor output varies inversely with distance from the radius. There are no abrupt transitions or bends in the curve as with the other two profile schemes. The sensor output drops to zero at a considerable distance from the object; when the robot is about to strike the object, the sensor output is maximum.

See also DISTANCE MEASUREMENT and PROXIMITY SENSING.

MANIPULATOR

A *manipulator* consists of a *robot arm*, and the *gripper* or *end effector* at the end of the arm. The term can also refer to a remotely controlled robot. See END EFFECTOR, ROBOT ARM and ROBOT GRIPPER.

MASTER–SLAVE MANIPULATOR

See TELEPRESENCE.

MEAN TIME BEFORE FAILURE/MEAN TIME BETWEEN FAILURES (MTBF)

The performance of a robot, computer, or other machine can be specified in various ways. Two of the most common are the *mean time before failure* and the *mean time between failures*, both abbreviated MTBF.

Component

For a single component, such as an *integrated circuit*, the MTBF (mean time before failure) is the length of time you can expect the device to work before it fails. This is found by testing a number of components and averaging how long they keep working.

A simplified example of MTBF, calculated in hours on the basis of the performance of five identical, hypothetical light bulbs, is shown in the drawing. The lifetimes are averaged to get the result. For the results to be meaningful, the number of samples must be much greater than five. Testing a large number of components, such as 1000 or even 10,000, eliminates coincidentally skewed results.

System

In the case of a system such as a robot or computer, the mean time between failures is determined according to how often the machine breaks down. As with the component-testing method, it is best to use many identical

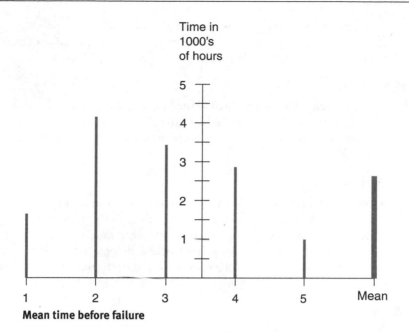

Time in
1000's
of hours

Mean time before failure

machines for this test. Conditions of the test should be as much like real life as possible.

With a large system, there are many components, any of which might malfunction. In general, the more complex the system, the shorter is the MTBF, if all other factors are held constant. This does not, of course, mean that simple systems are better than sophisticated ones. Therefore, for robots and computers, the MTBF is not a direct or overall indicator of the quality of the machine.

See also QUALITY ASSURANCE AND CONTROL.

MECHATRONICS

Mechatronics is a term that first appeared in Japan. It is a combination of the words "mechanics" and "electronics," and refers to the technology used in robotics. The term has the same literal meaning as *electro-mechanics*. In Japan, mechatronics became synonymous with industrial and economic power during the latter half of the twentieth century.

After World War II, the Japanese adopted the motto, "Catch and surpass the West." They hoped to do this by hard work, innovation, and devotion to quality: the things that have made the United States prosperous. Japan exports mechatronic equipment today. Sometimes it is called "Japan, Inc."

With the development of *artificial intelligence (AI)* in robotic systems, the importance of mechanics may diminish, and the importance of electronics can be expected to increase.

MEDICAL ROBOT

Medical science has become one of the largest industries in the civilized world. There are many possible ways in which robots might be used in this industry. The most likely scenario involves a mobile robotic attendant or nurse's assistant.

Paranurse

A *robotic paranurse* (nurse's assistant) might roll along on three or four wheels mounted in its base. Operation of the robot can be supervised from the nurses' station. This robot works very much like a *personal robot.* It can deliver meals, pick up the trays, and dispense medication in pill form. A robotic paranurse can also, at least in theory, take measurements of a patient's vital signs (temperature, heart rate, blood pressure, and respiration rate).

In the body

Robotic devices can be used as artificial limbs. This allows amputees and paraplegics to function almost as if they had never been injured.

A more radical notion conceived by some robotics researchers involves the manufacture of robot antibodies. These microscopic creatures could be injected into a patient's bloodstream, and would go around seeking out viruses or bacteria and destroying them. The robot controller could be programmed to cause the tiny machines to go after only a certain type of microorganism. Some researchers have suggested that organic compounds might be assembled molecule by molecule to make *biological robots.*

Entertainment

One of the more interesting applications of robots in medicine is to entertain the patient. This is where *artificial intelligence (AI)* becomes important. For children, robots can play simple games and read stories. For adults, robots can read aloud and carry on conversations.

The chief argument used against medical robots holds that sick people need a human touch, which a machine cannot provide. The counterargument asserts that medical robots are not intended to replace human doctors and nurses, but only to take up some of the slack, helping to relieve patients' boredom while freeing human beings to look after more critical things. Some patients, especially children, find robots immensely entertaining. Anyone who has stayed in a hospital for more than a couple of days knows how tedious it can become. A robot that can tell a few good jokes can brighten any patient's day.

See also PERSONAL ROBOT.

MEMORY

Memory refers to the storage of binary data in the form of high and low levels (logic ones and zeros). There are several forms of memory.

The amount of memory capacity is a factor in determining how "smart" a computer is. It is also helpful in choosing the right level of controller for a robotic system. Memory is measured in bytes, kilobytes (kB), megabytes (MB), gigabytes (GB), and terabytes (TB).

Random-access memory (RAM)

Random-access memory (RAM) chips store data in matrices called *arrays*. The data can be addressed (selected) from anywhere in the matrix. Data are easily changed and stored back in RAM, in whole or in part. A RAM is sometimes called a *read/write memory*.

An example of RAM is a word-processing computer file. This definition was written in semiconductor RAM, along with all the definitions of terms starting with the letter M, before being stored on disk, processed, and finally printed.

There are two kinds of RAM: *dynamic RAM* (DRAM) and *static RAM* (SRAM). DRAM employs *integrated-circuit* (IC) transistors and capacitors; data is stored as charges on the capacitors. The charge must be replenished frequently, or it will be lost to discharge. Replenishing is done several hundred times per second. SRAM uses a circuit called a *flip-flop* to store the data. This gets rid of the need for constant replenishing of charge, but the trade-off is that SRAM chips require more elements to store a given amount of data.

Volatile versus nonvolatile RAM

With any RAM, the data are erased when the appliance is switched off, unless some provision is made for *memory backup*. The most common means of memory backup is the use of a cell or battery. Modern IC memories need so little current to store their data that a backup battery lasts as long in the circuit as it would on the shelf. A new form of nonvolatile RAM, known as *flash memory*, can store large amounts of data indefinitely even with the power removed.

A memory that disappears when power is removed is called a *volatile memory*. If memory is retained when power is removed, it is *nonvolatile*.

Read-only memory (ROM and PROM)

By contrast to RAM, *read-only memory* (ROM) can be accessed, in whole or in part, but not written over. A standard ROM is programmed at the factory. This permanent programming is known as *firmware*. There are also ROM chips that a user can program and reprogram. This type of memory is known as *programmable read-only memory* (PROM).

Erasable PROM

An *erasable programmable read-only memory* (EPROM) chip is an IC whose memory is of the read-only type, but that can be reprogrammed by a certain procedure. It is more difficult to rewrite data in an EPROM than in a RAM; the usual process for erasure involves exposure to ultra-violet (UV) radiation. An EPROM chip can be recognized by the presence of a transparent window with a removable cover, through which the UV is focused to erase the data. The chip must be taken from the circuit in which it is used, exposed to the UV for several minutes, and then repro-grammed via a special process.

There are EPROMs that can be erased by electrical means. Such an IC is called an *electrically erasable programmable read-only memory* (EEPROM). These chips do not have to be removed from the circuit for reprogramming.

See also INTEGRATED CIRCUIT.

MEMORY ORGANIZATION PACKET

One of the most promising aspects of *artificial intelligence* (*AI*) is its use as a tool for predicting future events, based on what has happened in the past. This process is helped by arranging the computer memory into generalizations called *memory organization packets* (MOPs). Some crude examples of MOPs are the following statements:

- If the wind shifts to the east and the barometer falls, it will usually rain (or snow in the winter) within 24 hours.
- If the wind shifts to the west and the barometer rises, clearing will usually occur within a few hours.
- Light winds and a steady, high barometric pressure usually mean little weather change for at least 24 hours.
- Foul weather with a steady, low barometric pressure usually means bad weather for at least the next 24 hours.

These are broad generalizations, and they apply only in certain parts of the world (the temperate latitudes over land), but they are MOPs based on the experience of meteorologists over the past several centuries.

In AI, the system can be programmed to find the most valid MOPs based on available data. Then it can apply these MOPs in the most effective possible way to make a forecast in a given situation.

MESSAGE PASSING

Message passing refers to the repeated transfer of complex data among computers in an artificially intelligent system.

When a message is passed along many times, various things can happen to change the content (see illustration). Have you ever told someone a story, only to hear it sometime later in much different form? The same thing can happen in computer systems. Noise and distortion can alter signals, but this has largely been overcome by modern digital transmission methods. With sophisticated computers, another bugaboo crops up. An artificially intelligent system, designed to evaluate data subjectively rather than merely process it, can misinterpret a message, or even embellish it in ways the users did not intend and cannot predict.

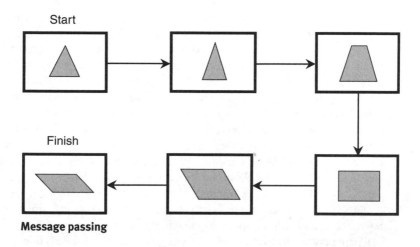

Message passing

METAL-OXIDE SEMICONDUCTOR (MOS)

The oxides of certain metals exhibit insulating and dielectric properties. So-called *metal-oxide-semiconductor* (MOS) electronic components have been in widespread use for years. MOS materials include compounds such as aluminum oxide and silicon dioxide. MOS devices are noted for their low power requirements. MOS *integrated circuits* (ICs) have high component density and high operating speed.

All MOS devices are subject to damage by the discharge of static electricity. Therefore, care must be exercised when working with MOS components. MOS ICs and transistors should be stored with the leads inserted into conducting foam, so large potential differences cannot develop. When building, testing, and servicing electronic equipment in which MOS devices are present, the technician's body and all test equipment should be kept at direct-current (DC) ground potential.

Metal-oxide-semiconductor technology lends itself well to the fabrication of digital ICs. Several MOS *logic families* have been developed. These are especially useful in high-density memory applications. Many microcomputer chips make use of MOS technology.

See also INTEGRATED CIRCUIT.

METRIC PATH PLANNING

Metric path planning is a scheme of robotic navigation in which the machine attempts to find the optimum path between two points. This generally requires a *computer map* of the environment, containing all possible (or probable) routes the robot might take from the starting point to the goal.

A common example of metric path planning is the choice of a route along a highway system between two towns. Suppose a traveler intends to take a trip from town X to town Y. These are the *initial node* and the *goal node,* respectively. A map is required, showing all the highways and roads between the two towns that constitute possible or reasonable routes. This map constitutes the *representation.* The more detail the map shows, the better. The map should be as up to date as possible, and should include such information as whether a road is two-lane or four-lane, zones where road maintenance is being carried out, the irregularity of the terrain, and the general traffic density likely to be found on each section of road. Using this information, the traveler plans the trip; this planning process constitutes an *algorithm.* Stop-overs might be planned, depending on the length of the trip; these are *intermediate nodes* or *waypoints.*

A robot can plan its route in exactly the same way as a traveler going from town X to town Y. Ideally, the machine will choose one, and only one, optimal path between its starting point and its goal. This optimal path might be the one that takes the least time; alternatively, it might be the one that requires the least expenditure of energy. The most time-efficient path might coincide with the most energy-efficient one, but this is not always the case.

See also COMPUTER MAP. Compare GRAPHICAL PATH PLANNING and TOPOLOGICAL PATH PLANNING.

MICROCOMPUTER

A *microcomputer* is a small computer with the *central processing unit* (CPU) enclosed in a single *integrated-circuit* (IC) package. The microcomputer CPU is sometimes called a *microprocessor.*

Microcomputers vary in sophistication and memory storage capacity, depending on the intended use. Some personal microcomputers are available for less than $100. Such devices employ liquid-crystal displays (LCDs) and have typewriter-style keyboards. Larger microcomputers are used by more serious computer hobbyists and by small businesses. Such microcomputers typically cost from several hundred to several thousand dollars.

Microcomputers are often used for the purpose of regulating the operation of electrical and electromechanical devices. This is known as *microcomputer control*. Microcomputer control makes it possible to perform complex tasks with a minimum of difficulty. Microcomputer control is widely used in such devices as robots, automobiles, and aircraft. For example, a microcomputer can be programmed to switch on an oven, heat the food to a prescribed temperature for a certain length of time, and then switch the oven off again. Microcomputers can be used to control automobile engines to enhance efficiency and gasoline mileage. Microcomputers can navigate and fly airplanes. It has been said that a modern jet aircraft is really a giant robot, because it can (in theory at least) complete a flight all by itself, without a single human being on board.

One of the most recent, and exciting, applications of microcomputer control is in the field of medical electronics. Microcomputers can be programmed to provide electrical impulses to control erratically functioning body organs, to move the muscles of paralyzed persons, and for various other purposes.

See also BIOMECHANISM and BIOMECHATRONICS.

MICROKNOWLEDGE

Microknowledge is detailed machine knowledge. In a smart robot or computer system, microknowledge includes logic rules, programs, and data stored in memory.

An example of microknowledge is the precise description of a person. In a personal smart robot, microknowledge allows the machine to recognize its owner(s). This microknowledge can, ideally, also let the robot know if a person approaching it is someone it has never met before. Another example of microknowledge is a *computer map* of the work environment. Compare MACROKNOWLEDGE.

MICROWAVE DATA TRANSMISSION

Microwave data transmission refers to the sending and receiving of wireless data at extremely high radio frequencies. Microwaves are very short electromagnetic waves, but they have longer wavelengths than infrared (IR) energy. Microwaves travel in essentially straight lines through the atmosphere, and are not affected by the ionosphere. Thus, they can easily pass from Earth's surface into space, and from space to the surface.

Microwaves are useful for short-range, high-reliability data links. Satellite communication and control are generally carried out at microwave frequencies. The microwave region contains a vast amount of spectrum space, and can hold many broadband signals.

Microwave radiation can cause heating of certain materials. This heating can be dangerous to human beings when the microwave radiation is intense. When working with microwave equipment, care must be exercised to avoid exposure to the rays.

The illustration shows a simplified block diagram of a microwave transmitter and receiver, including the antennas. The antennas are highly directional and must be aimed at each other with a line of sight between them. Microwave data transmission is useful in robotic systems in which a line of sight can be maintained between machines while they are in communication. Compare LASER DATA TRANSMISSION.

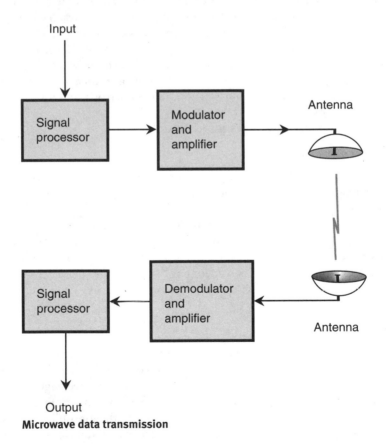

Microwave data transmission

MILITARY ROBOT

Whenever new technologies are developed, military experts look for ways in which the devices and systems can be used in warfare. *Military robots* have received special attention because, if machines could take the place of human beings in combat, there would be fewer human casualties. For example, *androids* could be used as robotic soldiers, as technicians, and for many other tasks humans would otherwise have to do. *Medical robots* could help in hospitals for humans who are injured. Robotic aircraft and tanks have existed for some time.

Artificial intelligence (AI) has also aroused the interest of military minds. With the aid of supercomputers, war strategy might be optimized. Computers can make decisions without being affected by emotions.

The following definitions concern robotic devices, computers and AI systems that have significant potential for use in the military: ADAPTIVE SUSPENSION VEHICLE, ANDROID, AUTONOMOUS ROBOT, BIOLOGICAL ROBOT, FLIGHT TELEROBOTIC SERVICER, FLYING EYEBALL, INSECT ROBOT, MEDICAL ROBOT, POLICE ROBOT, SECURITY ROBOT, SENTRY ROBOT, SUBMARINE ROBOT, TELEOPERATION, and TELEPRESENCE.

MODEM

The term *modem* is a contraction of *modulator/demodulator.* A modem interfaces a computer to a communications link, allowing robots to communicate with each other and/or with a central controller.

A computer works with binary *digital* signals, which are rapidly fluctuating direct currents. In order for digital data to be conveyed over a communications circuit, the data must usually be converted to *analog* form. This is done by changing the digit 1 into an audio tone, and the digit 0 into another tone with a different pitch. The result is an extremely fast back-and-forth alternation between the two tones. In *modulation*, digital data is changed into analog data. It is a type of *digital-to-analog (D/A) conversion. Demodulation* changes the analog signals back to digital ones; this is *analog-to-digital (A/D) conversion.*

Modems work at various speeds, usually measured in *bits per second (bps).* You will often hear about *kilobits per second (kbps),* where 1 kbps = 1000 bps, or *megabits per second (Mbps),* where 1 Mbps = 1000 kbps. The higher the bps figure, the faster the data are sent and received through the modem. Modems are rated according to the highest data speed they can handle.

The illustration is a functional block diagram of a modem. The modulator, or D/A converter, changes the digital computer data into audio tones. The demodulator, or A/D converter, changes the incoming audio tones into digital signals for the computer.

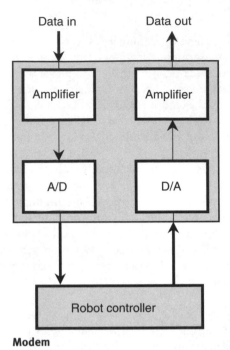

Modem

MODULAR CONSTRUCTION

A few decades ago, electronic equipment was constructed in a much different way than it is today. Components were mounted on tie strips, and wiring was done in point-to-point fashion. This kind of wiring is still used in some high-power radio transmitters, but in recent years, *modular construction* has become the rule.

In the modular method of construction, individual *circuit boards* are used, and each board (also called a *card*) has a specific function. Sometimes, several cards are combined into a *module*. The cards or modules are entirely removable, usually with a simple tool resembling pliers. Edge connectors facilitate easy replacement. The edge connectors are wired together for interconnection among cards and modules.

Modular construction has simplified the maintenance of complicated apparatus. In-the-field repair consists of identification, removal, and replacement of the faulty card or module. The defective unit is sent to a central facility, where it is repaired using sophisticated equipment. Once the card or module has been repaired, it can serve as a replacement unit when the need arises.

See also **INTEGRATED CIRCUIT.**

MODULUS

See NUMERATION.

MOS

See METAL-OXIDE SEMICONDUCTOR (MOS).

MOTOR

A *motor* converts electrical or chemical energy into mechanical energy. *Electric motors* are commonly used in robots. They can operate from alternating current (AC) or direct current (DC), and can turn at a wide range of speeds, usually measured in *revolutions per minute (rpm)* or *revolutions per second (rps)*. Motors range in size from the tiny devices in a wristwatch to huge, powerful machines that can pull a train.

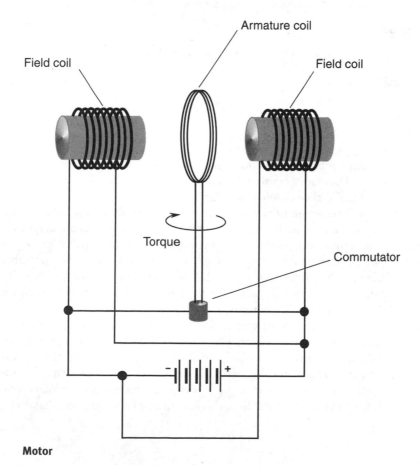

Motor

All motors operate by means of electromagnetic effects. Electric current flows through a set of coils, producing magnetic fields. The magnetic forces result in rotation. The greater the current in the coils, the greater is the rotating force. When the motor is connected to a load, the force needed to turn the shaft increases. The more the force, the greater the current flow, and the more power is drawn from the power source.

The illustration is a functional diagram of a typical electric motor. One set of coils rotates with the motor shaft. This is called the *armature coil*. The other set of coils is fixed, and is called the *field coil*. The *commutator* reverses the current with each half-rotation of the motor, so that the shaft keeps turning in the same direction.

The electric motor operates on the same principle as an electric generator. In fact, some motors can be used as generators.

See also GENERATOR, SELSYN, and SERVOMECHANISM.

MTBF

See MEAN TIME BEFORE FAILURE/MEAN TIME BETWEEN FAILURES (MTBF).

MULTIAGENT TEAM

See INSECT ROBOT.

MULTIPLEX

Multiplex is the transmission of two or more messages over the same line or channel at the same time. Multiplex transmission is done in various ways. The most common methods are *frequency-division multiplex (FDM)* and *time-division multiplex (TDM)*.

In FDM, a communications channel is broken down into subchannels. Suppose a channel is 24 kHz (kilohertz) wide. Then it can theoretically hold eight signals 3 kHz wide. The frequencies of the signals must be just right, so they do not overlap. Usually there is a little extra space on either side of each subchannel to ensure that overlapping does not occur. In FDM, data is transmitted in parallel form. That is, the information in all the channels is sent simultaneously.

Sometimes, data are cumbersome to transmit in parallel form. Such data can be converted to serial form using TDM. In this mode, signals are broken into pieces "timewise," and then the pieces are sent in a rotating sequence. This slows the rate of data transfer by a factor equal to the number of signals. For example, if each of six messages is 1 s long if sent by itself at full speed, the time-division-multiplexed signal will take 6 s.

N

NAND GATE

See LOGIC GATE.

NANOCHIP

Researchers are always striving to get more "computer power" into less physical space. This means superminiaturization of electronic components. This is especially important to the development of *artificial intelligence* (AI).

There is a practical limit to how many logic gates or switches can be etched onto an *integrated circuit* (IC), or chip, of a given size. This limit depends on the precision of the manufacturing process. As methods have improved, the density of logic gates on a single chip has increased. However, this can go only so far.

It has been suggested that, rather than etching the logic gates into silicon to make computer chips, engineers might approach the problem from the opposite point of view. Is it possible to build chips atom by atom? This process would result in the greatest possible number of logic gates or switches in a given volume of space. A hypothetical chip of this sort has been called a *nanochip*, because the individual switches have dimensions on the order of a few nanometers. One nanometer (1 nm) is 0.000000001 meter (10^{-9} m), or a millionth of a millimeter.

See also BIOCHIP and INTEGRATED CIRCUIT.

NANOROBOTICS

Superminiature robots, called *nanorobots*, might find all sorts of exotic applications. Roboticist *Eric Drexler* has suggested that such machines might serve as programmable antibodies, searching out and destroying harmful bacteria and viruses in the human body. In this way, diseases could be cured. The machines could also repair damaged cells.

Plagues that people once thought were eradicated for good, such as tuberculosis and malaria, are evolving new strains that resist conventional treatments. Biological research is largely a trial-and-error process. Suppose people could build millions of smart robots of nanometric dimensions, programmed to go after certain bacteria and viruses and kill them? Futurists believe this is possible. They envision building *molecular computers* from individual atoms of carbon, a fundamental ingredient of all living matter. These computers would store data in much the same way as does DNA, but the computers would be programmed by people rather than nature. These computers could be as small as 100 nm (10^{-7} m or 0.0001 mm) in diameter. Even an object this small has enough carbon atoms to make a chip with processing power equivalent to that of a typical personal computer.

There is a dark side to nanorobotics. Anything that can be used constructively can be used in some destructive way, as well. Programmable antibodies could, if they got into the hands of the wrong kinds of people, be used as biological weapons.

See also BIOCHIP and BIOMECHATRONICS.

NATURAL LANGUAGE

A *natural language* is a spoken or written language commonly used by people. Examples are English, Spanish, Russian, and Chinese.

In user-friendly computer and robotic systems, it is important that the machine be able to speak and/or write, and also to understand, as much natural language as possible. The more natural language a machine can accept and generate, the more people will be able to use the machine with less time spent learning how to do so.

Natural language will be extremely important in the future of hobby and personal robotics. If you want your robot *Cyberius* to get a cup of water, for example, you would like to tell it, "Cyberius, please get me a cup of water." You don't want to have to type a bunch of numbers, letters, and punctuation marks on a terminal, or speak in some arcane jargon that is nothing like normal talk.

See also SPEECH RECOGNITION and SPEECH SYNTHESIS.

NESTED LOOPS

In reasoning schemes or programs, logical *loops* are often found. A loop is a set of operations or steps that is repeated twice or more. Sometimes, loops occur inside other loops. They are then said to be *nested loops*.

The smaller loop in a nest usually involves fewer steps per repetition than the larger loop. The number of times a loop is followed is independent of the number of steps it contains. A small, secondary loop might be repeated 1 million times, while the larger loop surrounding it is repeated only 100 times.

The illustration is a flowchart that shows a simple example of nesting. Squares indicate procedural steps, such as "Multiply by 3 and then add 2." Diamonds are *IF/THEN/ELSE* steps, which are crucial to any loop. The question marks inside the diamonds indicate that a yes/no choice must be made, such as "Is *x* greater than 587?" The minus sign ($-$) is like a "No" to the question, in which case the process must go back to some earlier point. The plus sign ($+$) is like a "Yes," telling the process to go ahead.

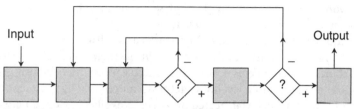

Nested loops

Nested loops are common in computer programs, especially when there are complicated mathematical calculations. Nesting of thought loops in the human mind probably also takes place. Any attempt at mapping human thought processes reveals fantastic, myriad twists and turns. Any attempt at modeling human thought requires the use of nested loops. This is a consideration in *artificial intelligence* (AI) research.

See also IF/THEN/ELSE.

NEURAL NETWORK

The term *neural network* refers to any of several forms of alternative computer technology. The basic idea behind all neural networks is to mimic the workings of the human brain.

Assets

Compared with digital computers, neural networks are fast. They can reach conclusions more rapidly than digital machines.

Neural networks are good at tasks such as *object recognition* and *speech recognition*. Neural networks can take small bits of information about an object, sound, or other complex thing, and fill in the gaps to get the whole. This was vividly demonstrated when an early version of a neural network took an incomplete (20 percent) radar image of a jet plane and, on the basis of that data, produced a complete graphic of the type of aircraft that caused the echoes.

Neural networks can learn from their mistakes, improving their performance after repeating a task many times. They also exhibit *graceful degradation,* so that if part of the system is destroyed, the rest can keep things going, albeit at a slower speed and/or with less accuracy.

Limitations

Neural networks are imprecise. If you ask one to balance your checkbook, it will come close, but it will not give an exact answer. Neural networks are not designed to do calculations of the sort a digital computer can carry out. A $5.00 calculator will outperform even the most complex neural network at basic arithmetic. In that sense, neural network technology resembles analog computer technology.

Another weakness of neural networks arises from the fact that they inevitably make mistakes as they zero in on their conclusions. Digital machines break problems down into miniscule pieces, meticulously grinding out a solution to a level of exactness limited only by the number of transistors that can be fabricated onto a chip of silicon. Neural networks tackle problems as a whole, modifying their outlook until the results satisfy certain conditions.

One might make the generalization that digital computers are analytical, while neural networks are intuitive.

Fuzzy logic

Digital machines recognize, at the fundamental level, two conditions or states: logic 1 and logic 0. These two logic states can be specified in terms of high/low, true/false, plus/minus, yes/no, red/green, up/down, front/back, or any other clear-cut dichotomy. The human brain is made up of neurons and synapses in a huge network, all of which can communicate with a vast number of others. In a neural network, "neurons" and "synapses" are the processing elements and the data paths between them. The earliest neural-network enthusiasts postulated that the human brain works like a huge digital machine, its neurons and synapses either "firing" or "staying quiet." Later, it was learned that things are more complicated than that.

In some neural networks, the neurons can send only two different types of signals, and represent the brain as theorized in the 1950s. However, results can be modified by giving some neurons and/or synapses more importance than others. This creates fuzzy logic, in which truth and falsity exist with varying validity.

Neural networks and artificial intelligence

Some researchers suggest that the ultimate goal of AI can be reached by a "marriage" of digital and neural-network technologies. Others think neural networks represent a dead end, and that digital technology has clearly proven itself to be the best way to build a computer. Neural-network research has gone through boom-and-bust cycles, partly as a result of differences of opinion.

Psychologists are interested in this technology because it might help them answer questions about the human brain. However, no neural network has come close to such complexity. Even the biggest neural networks conceived, with billions of neurons and trillions of synapses, would be less intelligent than a cat or dog. See OBJECT RECOGNITION, PATTERN RECOGNITION, SPEECH RECOGNITION, and SPEECH SYNTHESIS.

NODE

A *node* is a specific, important point in the path of a mobile robot or *end effector* as it navigates its environment. The starting point is called the *initial node*; the destination point is called the *goal node*. Decision points, if any, between the initial and goal nodes are *intermediate nodes*. In *metric path planning*, for example, a mobile robot in a complex environment with many obstructions navigates between the initial and goal nodes by first determining a set of intermediate nodes or *waypoints*, and then following the paths between those nodes.

In a communications network, the term *node* refers to a specific location at which data are processed or transferred. Examples include workstations, servers, printers, and cameras. In a robotic system, the individual robots constitute communications nodes if they can communicate with other robots or with a central controller. A central controller in a fleet of *insect robots* is a communications node.

See also COMPUTER MAP, GRAPHICAL PATH PLANNING, METRIC PATH PLANNING, and TOPOLOGICAL PATH PLANNING.

NOISE

Noise is a broadbanded alternating current (AC) or electromagnetic (EM) field. In contrast to signals, noise does not convey information. Noise can be natural or human-made.

Noise always degrades communications quality. It is a major concern in any device or system in which data are sent from one place to another, such as a fleet of mobile robots that must exchange data, or a swarm of *insect robots* under the supervision of a central controller. The higher the noise level, the stronger a signal must be if it is to be received error-free. At a given signal power level, higher noise levels translate into more errors and reduced communications range.

The illustration is a spectral display of signals and noise, with amplitude as a function of frequency. The background noise level is called the *noise floor*. The vertical lines, or pips, indicate signals that are stronger than the noise. Signals below the noise floor do not show up on the display, and cannot be received unless some means is found to reduce the noise level (that is, to lower the noise floor).

The noise level in a system can be minimized by using components that draw the least possible current. Noise can also be kept down by lowering the temperature of all the components in the system. Some experimentation has been done at extremely cold temperatures; this is called *cryogenic technology.*

The narrower the bandwidth of a signal, in general, the better the obtainable signal-to-noise ratio will be. However, this improvement takes place at the expense of data speed. Fiberoptic systems are relatively immune to noise effects. Digital transmission methods are superior to analog methods in terms of noise immunity. All these factors notwithstanding, there is a limit to how much the noise level can be reduced. Some noise will exist no matter what technology is used.

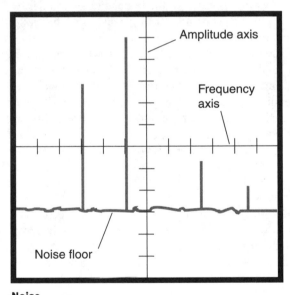

Noise

NONACTIVE COOPERATION

See **COOPERATION.**

NOR GATE

See **LOGIC GATE.**

NOT GATE

See **LOGIC GATE.**

NUCLEAR SERVICE ROBOT

Robots are well suited to handling dangerous materials. This is because, if there is an accident in an environment populated only by machines, no human lives will be lost. In the case of radioactive substances, robots can be used and operated by remote control, so that people will not be exposed to the radiation. The remote control is accomplished by means of *tele-operation* and/or *telepresence.*

Nuclear service robots have been used for some time in the maintenance of atomic power plants. One such machine, called ROSA, was designed and built by Westinghouse Corporation. It has been used to repair and replace heat-exchanger tubes in the boilers. The level of radiation is extremely high in this environment. It is difficult for human beings to do these tasks without endangering their health. If people spend more than a few minutes per month at such work, the accumulated radiation dose exceeds safety limits. Long-term overexposure to radiation increases the incidence of cancer and birth defects. Short-term, extreme exposure can cause radiation sickness or death.

Robots disarm nuclear warheads. If an errant missile is found with an unexploded warhead, it is better to use machines to eliminate the danger, rather than subjecting people to the risk (and mental stress) of the job.

See also **TELEOPERATION** and **TELEPRESENCE.**

NUMERATION

Several number systems are used in computer science, digital electronics, and robotics. The scheme most often used by people is *modulo 10,* also called the *decimal number system.*

Decimal numbers

The *decimal number system* is also called *modulo 10, base 10,* or *radix 10.* Digits are elements of the set {0, 1, 2, 3, 4, 5, 6, 7, 8, 9}. The digit immediately to the left of the radix (decimal) point is multiplied by 10^0, or 1. The next digit to the left is multiplied by 10^1, or 10. The power of 10 increases as you move farther to the left. The first digit to the right of the radix point is multiplied by a factor of 10^{-1}, or 1/10. The next digit to the right is multiplied by 10^{-2}, or 1/100. This continues as you go farther to the right. Once the process of multiplying each digit is completed, the resulting values are added. This is what is represented when you write a decimal number. For example,

$$2704.53816 = 2 \times 10^3 + 7 \times 10^2 + 0 \times 10^1 + 4 \times 10^0$$
$$+ 5 \times 10^{-1} + 3 \times 10^{-2} + 8 \times 10^{-3} + 1 \times 10^{-4} + 6 \times 10^{-5}$$

Binary numbers

The *binary number system* is a method of expressing numbers using only the digits 0 and 1. It is sometimes called *base 2, radix 2,* or *modulo 2.* The digit immediately to the left of the radix point is the "ones" digit. The next digit to the left is a "twos" digit; after that comes the "fours" digit. Moving farther to the left, the digits represent 8, 16, 32, 64, and so on, doubling every time. To the right of the radix point, the value of each digit is cut in half again and again, that is, 1/2, 1/4, 1/8, 1/16, 1/32, 1/64, and so on.

Consider an example using the decimal number 94:

$$94 = (4 \times 10^0) + (9 \times 10^1)$$

In the binary number system the breakdown is

$$1011110 = (0 \times 2^0) + (1 \times 2^1) + (1 \times 2^2)$$
$$+ (1 \times 2^3) + (1 \times 2^4) + (0 \times 2^5) + (1 \times 2^6)$$

When you work with a computer or calculator, you give it a decimal number that is converted into binary form. The computer or calculator does its operations with zeros and ones. When the process is complete, the machine converts the result back into decimal form for display.

Octal and hexadecimal numbers

Another numbering scheme is the *octal number system,* which has eight symbols, or 2^3. Every digit is an element of the set $\{0, 1, 2, 3, 4, 5, 6, 7\}$. Counting thus proceeds from 7 directly to 10, from 77 directly to 100, from 777 directly to 1000, and so on.

Yet another scheme, commonly used in computer practice, is the *hexadecimal number system,* so named because it has 16 symbols, or 2^4. These digits are the usual 0 through 9 plus six more, represented by A through F, the first six letters of the alphabet. The digit set is $\{0, 1, 2, 3, 4, 5, 6, 7, 8, 9, A, B, C, D, E, F\}$.

Comparison of values

The table compares values in modulos of 10 (decimal), 2 (binary), 8 (octal), and 16 (hexadecimal), from decimal 0 through decimal 64. In general, the larger the modulus, the smaller the numeral for a given value.

Numeration: comparison of values for decimal 0 through 64

Decimal	Binary	Octal	Hexadecimal
0	0	0	0
1	1	1	1
2	10	2	2
3	11	3	3
4	100	4	4
5	101	5	5
6	110	6	6
7	111	7	7
8	1000	10	8
9	1001	11	9
10	1010	12	A
11	1011	13	B
12	1100	14	C
13	1101	15	D
14	1110	16	E
15	1111	17	F
16	10000	20	10
17	10001	21	11
18	10010	22	12
19	10011	23	13
20	10100	24	14
21	10101	25	15
22	10110	26	16
23	10111	27	17
24	11000	30	18
25	11001	31	19
26	11010	32	1A
27	11011	33	1B
28	11100	34	1C
29	11101	35	1D
30	11110	36	1E
31	11111	37	1F

Numeration: comparison of values for decimal 0 through 64 (*Cont.*)

Decimal	Binary	Octal	Hexadecimal
32	100000	40	20
33	100001	41	21
34	100010	42	22
35	100011	43	23
36	100100	44	24
37	100101	45	25
38	100110	46	26
39	100111	47	27
40	101000	50	28
41	101001	51	29
42	101010	52	2A
43	101011	53	2B
44	101100	54	2C
45	101101	55	2D
46	101110	56	2E
47	101111	57	2F
48	110000	60	30
49	110001	61	31
50	110010	62	32
51	110011	63	33
52	110100	64	34
53	110101	65	35
54	110110	66	36
55	110111	67	37
56	111000	70	38
57	111001	71	39
58	111010	72	3A
59	111011	73	3B
60	111100	74	3C
61	111101	75	3D
62	111110	76	3E
63	111111	77	3F
64	1000000	100	40

OBJECT-ORIENTED GRAPHICS

One method by which a robotic vision system can define things is called *object-oriented graphics*, also known as *vector graphics*. This is a powerful technique that uses analog representations, rather than digital ones, to depict various shapes.

An example of an object-oriented graphic is a circle in the Cartesian coordinate plane, defined according to its algebraic equation. Consider the circle represented by the equation $x^2 + y^2 = 1$. This is called a *unit circle* because it has a radius of one unit, as shown in Fig. 1. The equation is easy for a computer to store in memory. Another, even simpler, rendition of this circle is its equation in polar coordinates (Fig. 2). In this system the unit circle is represented simply by $r = 1$. Both of these equations are mathematically exact representations of the circle, not digital approximations.

A digital, or bit-mapped, rendition of the unit circle requires approximation. The precision depends on the *image resolution*. An object-oriented representation is often more precise, and it avoids the problem of *jaggies*, also known as *aliasing*, that is always an artifact of a bit-mapped image. Compare BIT-MAPPED GRAPHICS.

See also COMPUTER MAP and VISION SYSTEM.

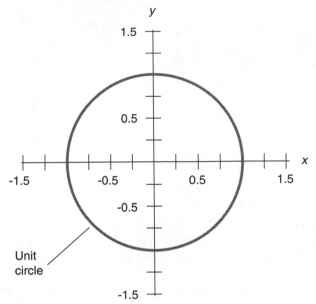

Object-oriented graphics—Fig. 1

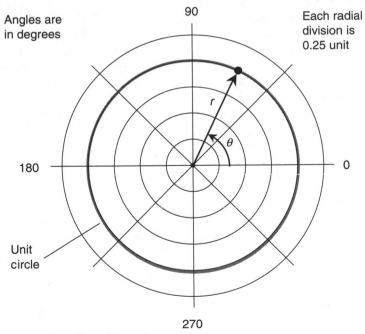

Object-oriented graphics—Fig. 2

OBJECT RECOGNITION

Object recognition refers to any method that a robot uses to pick something out from among other things. An example is getting a tumbler from a cupboard. It might require that the robot choose a specific object, such as "Jane's tumbler."

Suppose you ask your personal robot to go to the kitchen and get you a tumbler full of orange juice. The first thing the robot must do is find the kitchen. Then it must locate the cupboard containing the tumblers. How will the robot pick a tumbler, and not a plate or a bowl, from the cupboard? This is a form of *bin picking problem.*

One way for the robot to find a tumbler is with a *vision system* to identify it by shape. Another method is *tactile sensing.* The robot can double-check, after grabbing an object it thinks is a tumbler, to see whether it is cylindrical (the characteristic shape of a tumbler). If all the tumblers in your cupboard weigh the same, and if this weight is different from that of the plates or bowls, the robot can use weight to double-check that it has the right object. If a particular tumbler is required, then it will be necessary to have them marked in some way. *Bar coding* is a common scheme used for this purpose.

In general, the larger the number of characteristics that can be evaluated, the more accurate is the object recognition. Size, shape, mass (or weight), light reflectivity, light transmittivity, texture, and temperature are examples of variable characteristics typical of everyday objects.

See also BAR CODING, BIN PICKING PROBLEM, LOCAL FEATURE FOCUS, SENSOR FUSION, and VISION SYSTEM.

OCCUPANCY GRID

An *occupancy grid* is a graphical rendition of a robotic *sonar* or *radar* system. The region covered by the radar or sonar is broken up into squares, in the fashion of a rectangular coordinate system. Then each square is assigned a numerical value according to the probability of its being occupied. These values can range from -1 (100 percent certainty that the square is not occupied by an object) through 0 (equal chances that the square is occupied or unoccupied) to $+1$ (100 percent certainty that the square is occupied). Alternatively, the percentage probability of occupancy can be denoted, ranging from 0 (definitely not occupied) to 100 (definitely occupied).

When rendered as a two-dimensional (2-D) figure, an occupancy grid consists of a set of squares or rectangles, each with a number inside, representing the probability that the square is occupied. However, an occupancy grid can be rendered using colors instead of numbers if less precision is demanded. This technique is often used in meteorological radar or satellite displays showing intensity of precipitation, wind speed, temperatures of cloud tops, or other variables. Probabilities of occupancy might be assigned as follows: violet = 0–17%; blue = 18–33%; green = 34–50%; yellow = 51–67%; orange = 68–83%; red = 84–100%.

Grayscale 2-D renditions of occupancy grids are possible. An example is shown in the illustration, which represents a hypothetical work environment in which several birds' nests are located. This rendition has eight shades of gray. The darkest shades represent the highest probability that a bird is in a given sector at a given instant in time. White regions represent probabilities less than 1/8 (12.5 percent).

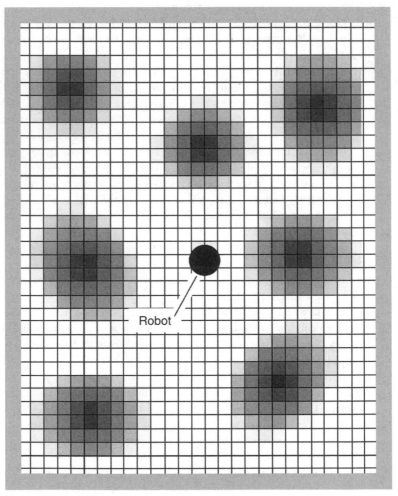

Occupancy grid

An occupancy grid can be rendered in three dimensions (3-D) by assigning the probabilities to values on an axis perpendicular to the plane of the grid itself. The regions of highest probability thus appear as "hills" or "mountains" while the regions of lowest probability appear as "valleys" or "canyons."

See also COMPUTER MAP, QUADTREE, RADAR, and SONAR.

OCTAL NUMBER SYSTEM

See NUMERATION.

OCTREE

See QUADTREE.

ODOMETRY

Odometry is a means of *position sensing*. It allows a robot to figure out where it is on the basis of two things: (1) a starting point, and (2) the motions it has made after having departed from that point.

Along a straight line, or in one dimension, odometry is performed by the mileage indicator in a car. The displacement, or distance traveled, is determined by counting the number of wheel rotations, based on a certain wheel radius. (If you switch to larger or smaller tires, you must realign the odometer in the car to get accurate readings.)

Distance traveled is equal to the integral of the speed over time. Graphically this can be represented by the area under a curve, as shown

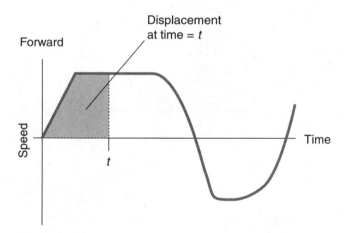

Odometry—Fig. 1

in Fig. 1. The displacement changes at a rate that depends on the speed. As long as you move forward, the displacement increases. If you go backward, the displacement decreases. Displacement can be either positive or negative with respect to the starting point.

In two dimensions, say in a room or over the surface of Earth, odometry is done by keeping constant track of velocity, which has components of both speed and direction.

Imagine boating in the open sea, starting from an island. You know the latitude and longitude of the island; you can measure your speed and direction constantly. You have a computer keep track of your speed and direction from moment to moment. Then, after any length of time, the computer can figure out where you are, based on past movements. It does this by integrating both components of velocity (speed and direction) simultaneously over time. Sailors know this as *ded reckoning* (short for *deductive reckoning*) of position.

A robot can use ded reckoning by having a microcomputer integrate its forward speed and its compass direction independently. This is called *double integration*. It is a rather sophisticated form of calculus, but a microcomputer can be programmed to do it easily. Figure 2 shows two-dimensional odometry based on speed and compass direction. The velocity vectors

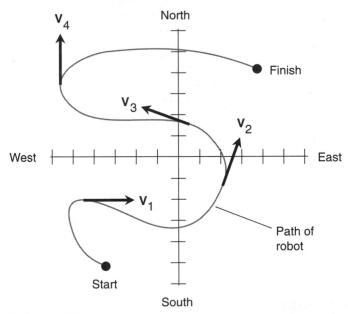

Odometry—Fig. 2

(V_1, V_2, V_3, and V_4 in this example) are constantly input to the micro-computer. The microcomputer "knows" the robot's coordinates at any moment, based on this information.

See also COMPUTER MAP and DISPLACEMENT TRANSDUCER.

OFFLOADING

It has been said that a machine is something that makes people's work easier. This is especially true of robots and smart computers. These devices can do many of the tedious or dangerous jobs that were done by humans in the past. As robotic technology advances, this process can be expected to continue. The replacement of human workers by robots and/or smart computers is called *robotization, automation,* or *computerization.* On a personal level, the use of robots and smart computers for mundane chores is called *offloading,* a term coined by futurist *Charles Lecht.*

According to Lecht, there is still much room for improvement in our lives. Even in our advanced, highly technological society, we spend time buying groceries, washing clothes, and vacuuming the floor. Creative people, in particular, often regard these things as a waste of time. But if the chores don't get done, people go hungry, wear dirty clothes, suffer with unsanitary living conditions. Some people hire servants to do their mundane chores, but not many people can afford a butler or a maid.

Robots can take over some routine daily chores. Lecht believes they eventually will, and that they will be affordable for almost everybody. This will free people to do more fun and creative things, Lecht says, such as paint pictures, write books, compose music, or play golf.

See also PERSONAL ROBOT.

OPEN-LOOP SYSTEM

The term *open-loop system* refers to any machine that does not incorporate a *servomechanism.* For this reason, open-loop robots are often called *nonservoed robots.* This type of robot depends, for its positioning accuracy, on the alignment and precision of its parts. There is no means for correcting positioning errors. The robot operates blind; it cannot compare its location or orientation with respect to its surroundings.

Open-loop systems can work faster than closed-loop, or servoed, robotic systems. This is because there is no feedback in an open-loop system, and therefore, no time is needed to process feedback signals and make positioning corrections. Open-loop systems are also less expensive than closed-loop systems. However, in tasks that require extreme accuracy, open-loop systems are not precise enough. This is especially true when a robot must make many programmed movements, one after the other. In some types of robotic systems, positioning errors accumulate unless they are corrected from time to time. Compare CLOSED-LOOP SYSTEM.

See also ERROR ACCUMULATION and SERVOMECHANISM.

OPERATIONAL AMPLIFIER

See INTEGRATED CIRCUIT.

OPTICAL CHARACTER RECOGNITION (OCR)

Computers can translate printed matter, such as the text on this page, into digital data. The data can then be used in the same way as if someone had typed it on a keyboard. This is done by means of *optical character recognition (OCR)*, a specialized form of *optical scanning*.

In OCR of printed matter, a thin laser beam moves across the page. White paper reflects light; black ink does not. The laser beam moves in the same way as the electron beam in a television camera or picture tube. The reflected beam is modulated; that is, its intensity changes. This modulation is translated by OCR software into digital code for use by the computer. In this way, a computer can actually "read" a magazine or book.

OCR is commonly used by writers, editors, and publishers to transfer printed data to digital media such as a computer's hard drive or CD-ROM (compact disc, read-only memory). Advanced OCR software can recognize mathematical symbols and other exotic notation, as well as capital and small letters, numbers, and punctuation marks.

Smart robots can incorporate OCR technology into their *vision systems*, enabling them to read labels and signs. The technology exists, for example, to build a smart robot with OCR that can get in a car and drive it anywhere. Perhaps someday, this will be commonly done. A robot owner might hand the robot a shopping list and say, "Please go get these things at the supermarket," and the robot will come back an hour later with the requested items.

For a robot to read something at a distance, such as a road sign, the image is observed with a video camera, rather than by reflecting a scanned laser beam off the surface. This video image is then translated by OCR software into digital data. See OBJECT RECOGNITION and VISION SYSTEM.

OPTICAL ENCODER

An *optical encoder* is an electronic device that measures the extent to which a mechanical shaft has rotated. It can also measure the rate of rotation (angular speed).

An optical encoder consists of a pair of *light-emitting diodes (LEDs)*, a *photodetector*, and a *chopping wheel*. The LEDs shine on the photodetector through the chopping wheel. The wheel has radial bands, alternately transparent and opaque (see illustration). The wheel is attached to the shaft. As the shaft turns, the light beam is interrupted. Each interruption actuates a counting circuit. The number of pulses is a direct function of the extent to which the shaft has rotated. The frequency of pulses is a direct function of the rotational speed. Two LEDs, placed in the correct positions, allow the encoder to indicate the direction (clockwise or counterclockwise) in which the shaft rotates.

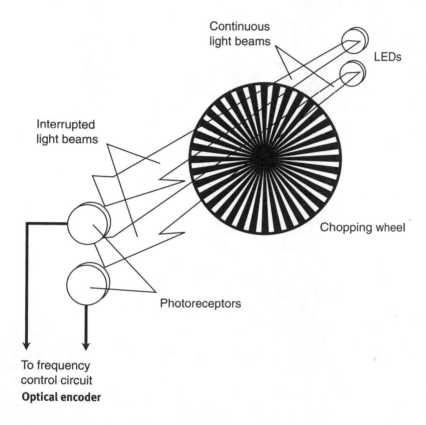

Continuous light beams

LEDs

Interrupted light beams

Chopping wheel

Photoreceptors

To frequency control circuit

Optical encoder

Optical encoders are used in various robotic applications. In particular, they are used in manipulators to measure the extent of joint rotation.

See also **ARTICULATED GEOMETRY** and **ROBOT ARM**.

OPTICAL INTERFEROMETER

See **PRESENCE SENSING**.

OPTIC FLOW

See EPIPOLAR NAVIGATION.

OR GATE

See LOGIC GATE.

ORIENTATION REGION

See LANDMARK.

P

PALLETIZING AND DEPALLETIZING

In manufacturing processes, it is often necessary to take objects from a conveyor belt and place them on a tray designed especially to fit them. The tray is called a *pallet,* and the process of filling it is called *palletizing.* The reverse process, in which objects are removed from the pallet and placed on the conveyor, is called *depalletizing.*

A complex sequence of motions is necessary to remove something from a conveyor, find an empty spot on a pallet, and place the object into the vacant spot correctly. Consider a pallet with holes for eight square pegs. One hole is filled; the other seven are vacant. Suppose that a robot is programmed to palletize pegs until the tray is full, then get another tray and fill it, and so on. Its instructions might be crudely depicted something like this:

1. Start palletizing routine.
2. Are pegs coming along the conveyor?
   a. If not, go to step 7.
   b. If so, go to step 3.
3. Is the pallet full?
   a. If not, keep it.
   b. If so, load it on the truck, get a new pallet, and put it in place to be filled up.
4. Get the first available peg from the conveyor.
5. Place the peg in the lowest numbered empty hole in the pallet.
6. Go to step 2.
7. Await further instructions.

PARADIGM

See **HIERARCHICAL PARADIGM, HYBRID DELIBERATIVE/REACTIVE PARADIGM,** and **REACTIVE PARADIGM.**

PARALLAX

Parallax is the effect that allows you to judge distances to objects and to perceive depth. Robots with *binocular machine vision* use parallax for the same purpose. The illustration shows the basic principle. Nearby objects appear displaced, relative to a distant background, when viewed with the left eye as compared to the view seen through the right eye. The extent of the displacement depends on the proportional difference between the distance to the nearby object and the distant reference scale, and also on the separation between the left eye and the right eye.

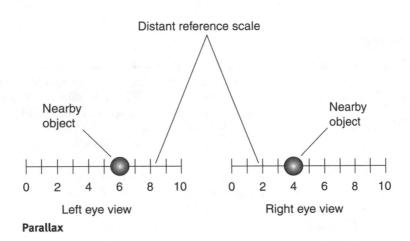

Distant reference scale

Nearby object   Nearby object

| 0 | 2 | 4 | 6 | 8 | 10 |

Left eye view   Right eye view

Parallax

Parallax can be used for navigation and guidance. If you are heading toward a point, that point seems stationary while other objects seem to move away from it. You see this while driving down a flat, straight highway. Signs, trees, and other road-side objects appear to move radially outward from a distant point on the road. A robot *vision system* can use this effect to sense the direction in which it is moving, its speed, and its location.

See also BINOCULAR MACHINE VISION.

PARALLEL DATA TRANSMISSION

See DATA CONVERSION.

PASSIVE TRANSPONDER

A *passive transponder* is a device that allows a robot to identify an object. *Bar coding* is an example. Magnetic labels, such as those on credit cards, automatic teller machine (ATM) bank cards, and retail merchandise are another example.

All passive transponders require the use of a sensor in the robot. The sensor decodes the information in the transponder. The data can be complex and the transponder can be tiny. In some systems, the information can be sensed from more than a meter away.

Suppose a robot needs to choose a drill bit for a certain application, and there are 150 bits in a tray, each containing a passive transponder with information about its diameter, hardness, recommended operating speeds, and its position in the tray. The robot can quickly select the best bit, install it, and use it. When the robot is done, the bit can be put back in its proper place.

See also BAR CODING.

PATH PLANNING

See GRAPHICAL PATH PLANNING, METRIC PATH PLANNING. and TOPOLOGICAL PATH PLANNING.

PATTERN RECOGNITION

In a robot vision system, one way to identify an object or decode data is by shape. *Bar coding* is a common example. Optical scanning is another. The machine recognizes combinations of shapes, and deduces their meanings using a microcomputer. In smart robots, the technology of pattern recognition is gaining importance. Researchers sometimes use *Bongard problems* to refine pattern-recognition systems.

Imagine a personal robot that you keep around the house. It might identify you because of combinations of features, such as your height, hair color, eye color, voice inflections, and voice accent. Perhaps your personal robot can instantly recognize your face, just as your friends do. This technology exists, but it requires considerable processing power and the cost is high. There are simpler means of identifying people.

Suppose your robot is programmed to shake hands with anyone who enters the house. In this way, the robot gets the fingerprints of the person. It has a set of authorized fingerprints in storage. If anyone refuses to shake hands, the robot can actuate a silent alarm to summon police robots. The same thing might happen if the robot does recognize the print of the person shaking its hand. This is a hypothetical and rather Orwellian scenario; many people would prefer not to enter a house so equipped. However, that fact in itself could arguably serve as a security enhancement.

See also BONGARD PROBLEM and OPTICAL CHARACTER RECOGNITION.

PERPENDICULAR FIELD

See POTENTIAL FIELD.

PERSONAL ROBOT

For centuries, people have imagined having a *personal robot.* Such a machine could be a sort of slave, not asking for pay (except maintenance expenses). Until the explosion of electronic technology, however, people's attempts at robot building resulted in clumsy masses of metal that did little or nothing of any real use.

Features

Personal robots can do all kinds of mundane chores around your house. Such robots are sometimes called *household robots.* Personal robots can be used in the office; these are called *service robots.* To be effective, personal robots must incorporate features such as *speech recognition, speech synthesis, object recognition,* and a *vision system.* Household robot duties might include:

- Car washing
- General cleaning
- Companionship
- Cooking
- Dishwashing
- Fire protection
- Floor cleaning
- Grocery shopping
- Intrusion detection
- Laundry
- Lawn mowing
- Maintenance
- Meal serving
- Child's playmate
- Snow removal
- Toilet cleaning
- Window washing

Around the office, a service robot might do things such as:

- Bookkeeping
- General cleaning
- Coffee preparation and serving
- Delivery
- Dictation
- Equipment maintenance
- Filing documents
- Fire protection

- Floor cleaning
- Greeting visitors
- Intrusion detection
- Meal preparation
- Photocopying
- Telephone answering
- Toilet cleaning
- Typing
- Window washing

Practical robots versus toys

Some personal robots have been designed and sold, but until recently, they were not sophisticated enough to be of any practical benefit. Most such robots are more appropriately called *hobby robots*. A good household robot, capable of doing even a few of the above chores efficiently and reliably, is beyond the financial means of ordinary people. As technology improves and becomes less expensive, the cost (in terms of a person's real earning power) will go down.

Simpler machines make good toys for children. Interestingly, if a robot is designed and intended as a toy, it often sells better than if it is advertised as a practical machine.

Questions and concerns

Robots must be safe to live around, and not pose any hazard to their owners, especially children. This can be ensured with good design. All robots should function in accordance with *Asimov's three laws*.

Suppose a practical personal robot were available for about the same price as a fine automobile. Would many people buy it? This is hard to predict. As boring as some of the above-mentioned tasks might seem, plenty of people enjoy doing them. Lawn mowing and snow removal can be good exercise. Lots of people like to cook. Some people will never entrust a robot to do things right, no matter how efficient and sophisticated the machines might be. Some people might prefer to save or invest money they could spend on a personal robot. Related definitions in this book include ASIMOV'S THREE LAWS, EDUCATIONAL ROBOT, FIRE-PROTECTION ROBOT, FOOD SERVICE ROBOT, GARDENING AND GROUNDSKEEPING ROBOT, HOBBY ROBOT, MEDICAL ROBOT, OFFLOADING, ROBOT CLASSIFICATION, ROBOT GENERATIONS, SECURITY ROBOT, SMART HOME, SPEECH RECOGNITION, and SPEECH SYNTHESIS.

PHONEME

A *phoneme* is an individual sound or syllable you make when you talk. Examples are "ssss," "oooo," and "ffff."

A voice can be displayed on an oscilloscope screen. The hardware is simple: a microphone, audio amplifier, and oscilloscope. When a person speaks into the microphone, a jumble dances across the screen. Phonemes look simpler than ordinary speech. Any waveform, no matter how complex, can be recognized or generated by electronic circuits. A speech synthesizer can, in theory, be made to sound exactly like anyone's voice, saying anything, with any inflection desired. The output of such a machine has precisely the same waveform, as seen on an oscilloscope, as the particular speaker's voice.

See also SPEECH RECOGNITION and SPEECH SYNTHESIS.

PHOTOELECTRIC PROXIMITY SENSOR

Reflected light can provide a way for a robot to tell if it is approaching something. A *photoelectric proximity sensor* uses a modulated light-beam generator, a photodetector, a frequency-sensitive amplifier, and a microcomputer. The illustration shows the principle of this device.

The light beam reflects from the object and is picked up by the photodetector. The light beam is modulated at a certain frequency, say 1000 Hz (hertz), and the detector has an amplifier that responds only to light modulated at that frequency. This prevents false imaging that might otherwise be caused by stray illumination such as lamps or sunlight. If the robot is approaching an object, the microcomputer senses that the reflected beam is getting stronger. The robot can then steer clear of the object.

This method of proximity sensing does not work for black or very dark objects, or for flat windows or mirrors approached at a sharp angle. These sorts of objects fool this system, because the light beam is not reflected back toward the photodetector.

See also ELECTRIC EYE and PROXIMITY SENSING.

PICTURE SIGNAL

See COMPOSITE VIDEO SIGNAL.

PIEZOELECTRIC TRANSDUCER

A *piezoelectric transducer* is a device that can convert acoustic waves to electrical impulses, or vice versa. It consists of a crystal, such as quartz or ceramic material, sandwiched between two metal plates, as shown in the illustration on page 226.

When an acoustic wave strikes one or both of the plates, the metal vibrates. This vibration is transferred to the crystal. The crystal generates weak electric currents when it is subjected to mechanical stress. Therefore, an alternating-current (AC) voltage develops between the two metal plates, with a waveform similar to that of the sound waves.

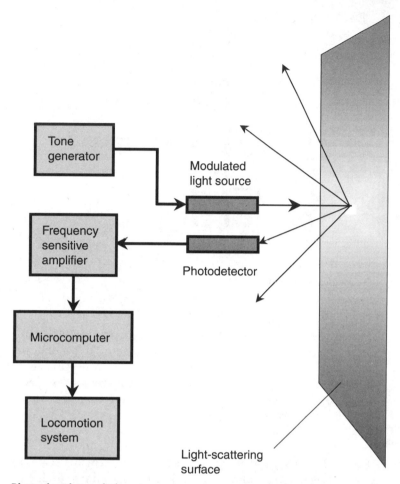

Photoelectric proximity sensor

If an AC signal is applied to the plates, it causes the crystal to vibrate "in sync" with the current. The result is that the metal plates vibrate also, producing an acoustic disturbance.

Piezoelectric transducers are common in ultrasonic applications, such as intrusion detectors and alarms. They are useful in robotic systems because they are small, light in weight, and demand little current for their operation. They are sensitive and can function underwater. Compare DYNAMIC TRANSDUCER and ELECTROSTATIC TRANSDUCER.

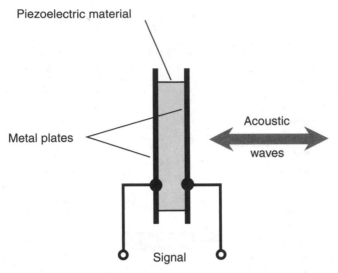

Piezoelectric transducer

PITCH

Pitch is one of three types of motion that a robotic *end effector* can make. It also refers to changes in the attitude (orientation) of a mobile robot in three dimensions. Pitch is generally an up-and-down variable. Extend your arm out straight, and point at something with your index finger. Then move your wrist so that your index finger points up and down along a vertical line. This motion is pitch in your wrist. Compare ROLL and YAW.

PIXEL

Pixel is an acronym that means "picture (pix) element." A pixel is the smallest region in a two-dimensional (2-D) video image or display. In a *composite video signal,* a pixel is the smallest unit that conveys information. These pixels are sometimes, but not always, coincident with the pixels on the display at the receiving end of the circuit.

If you look through a magnifying glass close-up at a television screen or computer monitor, you can see thousands of little dots. These are the pixels of the television or monitor screen itself. (Caution: Wear ultraviolet-protective sunglasses if you try this experiment.) In a grayscale image, each pixel is assigned a specific brilliance. In a color image, each pixel is assigned a primary color (red, green, or blue) and a specific brilliance.

The size of a pixel is important in robotic vision systems, because this determines the ultimate *image resolution*—how much detail the robot can see. The smaller the pixels, the better is the resolution. High-resolution robot vision requires better cameras, greater signal bandwidth, and more memory than low-resolution robot vision. High-resolution systems also cost more than low-resolution systems.

See also COMPOSITE VIDEO SIGNAL, RESOLUTION, and VISION SYSTEM.

PLAN/SENSE/ACT

See HIERARCHICAL PARADIGM, HYBRID DELIBERATIVE/REACTIVE PARADIGM, and REACTIVE PARADIGM.

PNEUMATIC DRIVE

A *pneumatic drive* is a method of providing movement to a robot manipulator. It uses compressed gas, such as air, to transfer forces to various joints, telescoping sections, and end effectors.

The pneumatic drive consists of a power supply, one or more motors, a set of pistons and valves, and a feedback loop. The valves and pistons control the movement of the gas. Because the gas is compressible, the drive cannot impart large forces without significant positioning errors. A feedback loop consists of one or more force sensors that can provide error correction and help the manipulator follow its intended path.

Pneumatically driven manipulators are used when precision and speed are not critical. Compare HYDRAULIC DRIVE.

POINT-TO-POINT MOTION

Some robot arms move continuously, and can stop at any point along the path. Others are able to stop only in specific places. When the *end effector* of

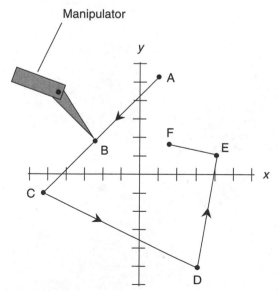

Point-to-point motion

a robot arm can attain only certain positions, the manipulator is said to employ *point-to-point motion.* The illustration shows point-to-point motion in which six stopping points, called *via points,* are possible (A through F).

In some robots that use point-to-point motion, the controller stores a large number of via points within the work envelope of the manipulator. These points are so close together that the resulting motion is continuous for practical purposes. Small time increments are used, such as 0.01 s or 0.001 s. This scheme is the robot-motion analog of bit-map computer graphics. Compare CONTINUOUS-PATH MOTION.

POLAR COORDINATE GEOMETRY

Industrial robot arms can move in various different ways, depending on their intended use. *Polar coordinate geometry* is a common two-dimensional

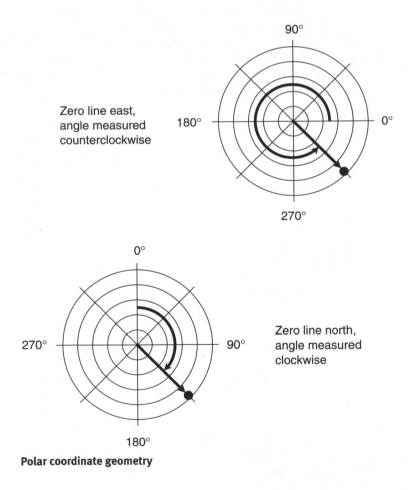

Polar coordinate geometry

(2-D) system. This term comes from the polar graph for mathematical functions. The drawings show standard polar coordinate systems.

The *independent variable* is the angle, in degrees or radians, relative to a defined zero line. There are two common methods of specifying the angle. If the zero line runs toward the right ("east"), then the angle is measured counterclockwise from it. If the zero line runs upward ("north"), then the angle is measured clockwise from it. The first scheme is common for mathematical displays and some robotic manipulators. The second method is used when the angle is a compass bearing or azimuth, as in navigation systems.

The *dependent variable* is the radius, or distance from the center of the graph. The units are usually all the same size in a given coordinate plot (for example, millimeters). In some cases, a logarithmic radius scale is used. This is often done when plotting transducer directional patterns. Compare ARTICULATED GEOMETRY, CARTESIAN COORDINATE GEOMETRY, CYLINDRICAL COORDINATE GEOMETRY, REVOLUTE GEOMETRY, and SPHERICAL COORDINATE GEOMETRY.

POLICE ROBOT

Can you imagine metal-and-silicon police officers, 2 m tall, capable of lifting whole cars with one arm and, at the same time, shooting 100 bullets per second from an end effector on the other arm? These types of *police robots* have been depicted in fiction. The technology to build such a machine exists right now. However, when and if robotic police machines are developed on a large scale in real life, they will probably be less sensational.

Police officers are often exposed to danger. If a remote-controlled robot could be used for any of the dangerous jobs that cops face, lives could be saved. This is the rationale for deploying robots in place of human officers. A robot police officer might work something like a robot soldier or drone. It could be teleoperated, with a human operator stationed in a central location, not exposed to risk. A mechanical cop could certainly be made far stronger than any human being. In addition, a machine has no fear of death, and can take risks that people might back away from.

Humans can maneuver physically in ways that no machine can match. A clever crook could probably elude almost any individual robot cop. Agility will be a key concern if a robot police officer is ever to apprehend anybody. Sheer force of numbers might overcome this problem. Perhaps a large swarm of small *insect robot* cops, strategically deployed, could track and catch a fleeing suspect.

Sophisticated, autonomous robot police officers might not prove cost-effective. A human operator must be paid to sit and teleoperate a robot. The robot itself will cost money to build and maintain, and if necessary, to repair or replace. People who roboticize a police force will have to weigh the saving of life against increased costs. Perhaps the cost of robotic technology will decline, and the quality will increase, until someday, part or most of our metropolitan police forces can be roboticized at a reasonable cost.

See also MILITARY ROBOT, SENTRY ROBOT, and SECURITY ROBOT.

POLYMORPHIC ROBOT

A *polymorphic robot,* also called a *shape-shifting robot,* is designed to conform to its environment by altering its geometry. There are numerous designs that can accomplish this. A simple example of a polymorphic robot gripper is the *active chord mechanism,* which conforms to objects by wrapping around them.

Specialized track-driven robots can change their shapes in order to travel over rough terrain, or climb and descend stairs. Such robots can also alter their body orientation (horizontal or vertical). Some robots are shaped like snakes, with numerous joints that allow them to maneuver in, and reach into, complex work spaces.

See also ACTIVE CHORD MECHANISM, TRACK-DRIVE LOCOMOTION, and TRI-STAR WHEEL LOCOMOTION.

POSITION SENSING

Robot *position sensing* falls into either of two categories. In the larger sense, the robot can locate itself. This is important in guidance and navigation. In the smaller sense, a part of a robot can move to a spot within its work envelope, using devices that tell it exactly where it is. Specific definitions in this book that deal with position sensing include CARTESIAN COORDINATE GEOMETRY, COMPUTER MAP, CYLINDRICAL COORDINATE GEOMETRY, DIRECTION FINDING, DIRECTION RESOLUTION, DISPLACEMENT TRANSDUCER, DISTANCE MEASUREMENT, DISTANCE RESOLUTION, EDGE DETECTION, EPIPOLAR NAVIGATION, EYE-IN-HAND SYSTEM, GUIDANCE SYSTEM, LANDMARK, LOCAL FEATURE FOCUS, ODOMETRY, PARALLAX, PHOTOELECTRIC PROXIMITY SENSOR, POLAR COORDINATE GEOMETRY, PROXIMITY SENSING, SONAR, SPHERICAL COORDINATE GEOMETRY, and VISION SYSTEM.

POTENTIAL FIELD

A *potential field* is a rendition of robot behavior or characteristics within a specific work area. Such fields are commonly rendered as vector arrays in a two-dimensional (2-D) coordinate system. The vectors can represent any quantity that affects the robot, or that the robot exhibits, such as magnetic field strength, velocity, or acceleration. More complex potential fields exist in three-dimensional (3-D) space. The following examples, and the accompanying illustrations, involve 2-D space (a flat surface) for simplicity.

Uniform

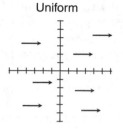

Attractive radial

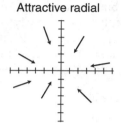

Repulsive radial

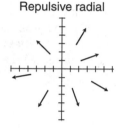

Tangential

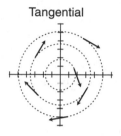

Perpendicular

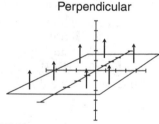

Potential field

Uniform field

In a *uniform potential field*, the vectors all point in the same direction and have the same magnitude, regardless of the location of the robot. All of the vectors point parallel to the work surface. An example of this sort of field is a steady wind acting on a robot. Another example is Earth's magnetic field in a work space that covers only a small part of the surface of the planet (for example, a few square kilometers), and which is located near the geomagnetic equator where the lines of flux are parallel to the surface.

Attractive radial field

An *attractive radial potential field* contains vectors that all point inward toward the origin, or center point, represented by (0, 0) in the Cartesian coordinate system. The vector magnitude might depend on the distance from the origin, but not necessarily. An example of an attractive radial field is a mapping of the force that exists as a robot carrying an electrically positive charge operates in the vicinity of an object carrying an electrically negative charge. In this case, the vector intensity increases as the distance between the robot and the origin decreases.

Repulsive radial field

A *repulsive radial potential field* contains vectors that all point outward away from the origin. As with the attractive field, vector magnitude might depend on the distance from the origin, but not necessarily. An example of a repulsive radial field is a mapping of the force that exists if a robot carrying an electrically positive charge operates in the vicinity of an object carrying an electrically positive charge (that is, the same polarity as the robot). In this case, the vector intensity increases as the distance between the robot and the origin decreases.

Tangential field

A *tangential potential field* contains vectors that point either clockwise or counterclockwise in concentric circles around the origin. The vector magnitude might vary depending on the distance from the origin, but not necessarily. An example of this type of field is the wind circulation surrounding an intense tropical hurricane. Another example is magnetic flux surrounding a straight wire carrying a steady, direct current, when the wire passes through the 2-D work surface at a right angle. In both of these cases, the vector intensity increases as the distance between the robot and the origin decreases.

Perpendicular field

In a *perpendicular potential field,* also called an *orthogonal potential field,* the vectors all point in the same direction and have the same magnitude, regardless of the location of the robot. All of the vectors point at a right angle to the work surface. An example of this sort of field is Earth's magnetic field in the immediate vicinity of either geomagnetic pole. Another example is the mapping of the force that occurs if a robot carrying an electric charge operates on a work surface that also carries an electric charge. If the robot and the surface have like charges, the force is repulsive (the vectors all point straight up); if the robot and the surface have opposite charges, the force is attractive (the vectors all point straight down).

See also COMPUTER MAP.

POWER SUPPLY

A *power supply* is a circuit that provides an electronic device with the voltage and current it needs for proper operation. The power from a typical utility outlet consists of alternating current (AC) at about 117 V. Most electronic equipment requires direct current (DC). Figure 1 is a block diagram showing the stages in a typical DC power supply. The stages include a *transformer,* a *rectifier,* a *filter,* and a *voltage regulator.*

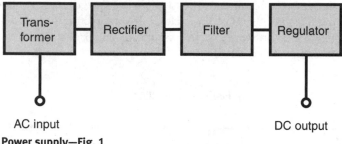

AC input   DC output

Power supply—Fig. 1

The transformer

Power-supply transformers are available in two types: the *step-down transformer* that converts AC to a lower voltage, and the *step-up transformer* that converts AC to a higher voltage. These are illustrated schematically in Fig. 2.

Most solid-state electronic devices, such as robot controllers and small robot motors, need only a few volts. The power supplies for such equipment use step-down power transformers. The physical size of the transformer depends on the current.

Step-down transformer

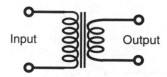

Input   Output

Step-up transformer

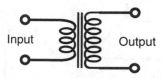

Input   Output

Power supply—Fig. 2

Some circuits require high voltage (more than 117 V DC). A cathode-ray tube (CRT) video display needs several hundred volts. The transformers in these appliances are step-up types.

The rectifier

The simplest rectifier circuit, called a *half-wave rectifier,* uses one diode to "chop off" half of the AC input cycle. Half-wave rectification is useful in supplies that do not have to deliver much current, or that do not need to be especially well regulated.

Half-wave rectifier

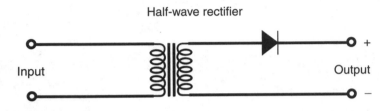

Full-wave center-tap rectifier

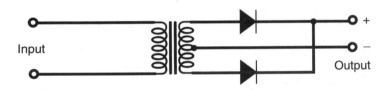

Full-wave bridge rectifier

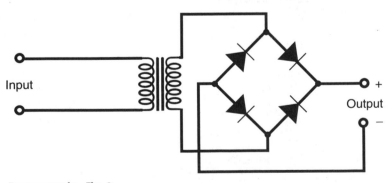

Power supply—Fig. 3

For high-current equipment, a *full-wave rectifier* is preferred. The full-wave scheme is also better when good voltage regulation is needed. This circuit makes use of both halves of the AC cycle to derive its DC output. There are two basic circuits for the full-wave supply. One version uses a center tap in the transformer, and needs two diodes. The other circuit uses four diodes and does not require a center-tapped transformer.

The half-wave, full-wave center-tap, and bridge rectifier circuits are shown schematically in Fig. 3.

The filter

Electronic equipment generally does not function well with the pulsating DC that comes straight from a rectifier. The ripple in the waveform must be smoothed out, so that pure, battery-like DC is supplied. A filter circuit does this.

The simplest possible filter is one or more large-value capacitors, connected in parallel with the rectifier output. Electrolytic or tantalum capacitors are used. Sometimes a large-value coil, called a *filter choke,* is connected in series, in addition to the capacitor in parallel. This provides a smoother DC output than the capacitor by itself. Two examples of inductance/capacitance filters are shown in Fig. 4.

Capacitor-input filter

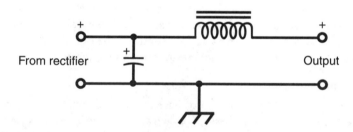

Choke-input filter

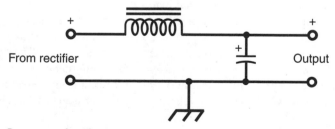

Power supply—Fig. 4

235

The voltage regulator

If a special kind of diode, called a *Zener diode*, is connected in parallel with the output of a power supply, the diode will limit the output voltage of the supply by as long as the diode has a high enough power rating. The limiting voltage depends on the particular Zener diode used. There are Zener diodes to fit any reasonable power-supply voltage.

When a power supply must deliver high current, a *power transistor* is used along with the Zener diode to obtain regulation. A circuit diagram of such a scheme is shown in Fig. 5.

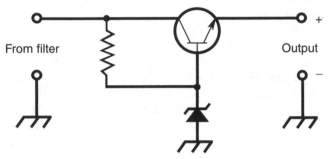

Power supply—Fig. 5

In recent years, voltage regulators have become available in *integrated-circuit* (IC) form. Such an IC, sometimes along with some external components, is installed in the power-supply circuit at the output of the filter. This provides excellent regulation at low and moderate voltages.

Transients and surges

The AC on the utility line does not have a clean, perfect, constant wave shape. Sometimes there are "spikes" known as *transients*. These last for only a few microseconds, but they can reach peak values of more than 1000 V. *Power surges* can also present a problem. In a surge, the voltage rises somewhat above normal for a half-second or so. Without some protection against the effects of transients and surges, sensitive electronic equipment, such as robot controllers, can malfunction.

The simplest way to get rid of most transients and surges is to use a commercially made *transient suppressor*, also called a *surge suppressor*. A more sophisticated power-processing device is an *uninterruptible power supply* (UPS). These are recommended for serious computer users, because they can prevent problems that would otherwise result from undervoltage and blackouts, as well as eliminating the effects of transients and surges.

Fuses and breakers

If a *fuse* blows, it must be replaced with another of the same rating. If the replacement fuse is rated too low in current, it will probably blow out right away, or soon after it has been installed. If the replacement fuse is rated too high in current, it might not protect the equipment.

Circuit breakers do the same thing as fuses, except that a breaker can be reset by turning off the power supply, waiting a moment, and then pressing a button or flipping a switch. Some breakers reset automatically when the equipment has been shut off for a certain length of time.

Safety issues

Power supplies can be dangerous. This is especially true of high-voltage circuits, but anything over 12 V should be treated as potentially lethal. In all AC-operated electronic apparatus, high voltage exists at the input to the supply (where 117 V appears). A CRT display has high voltages that operate its deflecting coils.

A power supply is not necessarily safe after it has been switched off. Filter capacitors hold a charge for a long time. In high-voltage supplies of good design, *bleeder resistors* are connected across each filter capacitor, so the capacitors will discharge in a few minutes after the supply is turned off. But don't bet your life on components that might not exist in a piece of hardware, and that can sometimes fail even when they are provided. If you have any doubt about your ability to repair a power supply, leave it to a professional.

See also ELECTROCHEMICAL POWER and SOLAR POWER.

PRESENCE SENSING

Presence sensing is the ability of a robot or other machine to detect the introduction of an object into the environment. Such a device can make use of bumpers, whiskers, visible light, infrared (IR), or acoustic sensors.

Bumpers and whiskers

The simplest presence sensors operate by direct physical contact. Their output is zero until they actually hit something. Then the output rises abruptly. *Bumpers* and *whiskers* work this way.

A bumper might be completely passive, making the robot bounce away from things that it hits. More often, a bumper has a switch that closes when it makes contact, sending a signal to the controller causing the robot to back away.

When whiskers hit something, they vibrate. This can be detected, and a signal sent to the robot controller. Whiskers might seem primitive, but they are a cheap and effective method to keep a machine from crashing into obstructions.

Electric eye

Another simple scheme for presence detection is an *electric eye*. Beams of IR or visible light are shone across points of entry such as doorways and window openings. Photodetectors receive energy from the beams. If any photodetector stops receiving its beam, a signal is generated. See ELECTRIC EYE.

Optical, IR, or microwave reflection sensor

An *optical presence sensor* is a device similar to the electric eye, except that it senses light beams reflected from objects, rather than interrupted by them. An *IR presence sensor* uses IR rather than visible light; a *microwave presence sensor* employs electromagnetic waves having short wavelengths (of the order of a few centimeters or less).

Beams of visible, IR, or microwave energy are shone into the work environment from various strategic locations. If any new object is introduced, and if it has significant reflectivity, the photodetectors will sense the reflected energy and cause a signal to be generated. The visible or IR system can be fooled by nonreflective objects. A good example is a robot coated with solid, flat black paint. Microwave systems may not respond to objects comprised entirely of nonconducting (dielectric) materials such as plastic or wood.

Optical, IR, or microwave interferometer

An *interferometer* can be used by a robot to detect the presence of an object or barrier at close range. It works based on wave interference, and can operate at any electromagnetic (EM) wavelength. Usually, EM energy in the radio microwave, IR, or visible range is used. When an object containing sufficiently reflective material intrudes into the work space, the reflected wave combines with the incident wave to generate an *interference pattern*. This wave interference can be detected, and sent to the robot controller.

The effectiveness of the interferometer depends on how well the object or barrier reflects energy at the wavelength used by the device. For example, a white-painted wall is more easily detected by an optical interferometer than a similar wall painted dull black. In general, an interferometer works better as the distance decreases, and less well as the distance increases. The amount of radio, IR, or optical *noise* in the robot's work environment is also important. The higher the noise level, the more limited is the range over which the sensor functions, and the more likely are false positives or negatives.

IR motion detector

A common presence-detection system employs an *IR motion detector*. Two or three wide-angle IR pulses are transmitted at regular intervals; these pulses cover most of the zone for which the device is installed. A receiving transducer picks up the returned IR energy, normally reflected from walls, floors, ceilings, and furniture. The intensity of the received pulses is noted by a microprocessor. If anything in the room changes position, or if a new object appears, there is a change in the intensity of the received energy. The microprocessor notices this change, and generates a signal. These devices consume very little power in regular operation, so batteries can serve as the power source.

Radiant heat detector

Infrared devices can detect changes in the indoor environment via direct sensing of the IR energy (often called *radiant heat*) emanating from objects. Humans, and all warm-blooded animals, emit IR. So does fire. A simple IR sensor, in conjunction with a microprocessor, can detect rapid or large increases in the amount of radiant heat in a room. The time threshold can be set so that gradual or small changes, such as might be caused by sunshine, will not trigger the signal, while significant changes, such as a person entering the room, will. The temperature-change (increment) threshold can be set so that a small animal will not actuate the alarm, while a full-grown person will. This type of device, like the IR motion detector, can operate from batteries.

The main limitation of radiant-heat detectors is the fact that they can be fooled. False alarms are a risk; the sun might suddenly shine directly on the sensor and trigger a presence signal. It is also possible that a person clad in a winter parka, boots, hood, and face mask, just entering from a subzero outdoor environment, might fail to generate a signal. For this reason, radiant-heat sensors are used more often as fire-alarm actuators than as presence detectors.

Ultrasonic motion detector

Motion in a room can be detected by sensing the changes in the relative phase of acoustic waves. An *ultrasonic motion detector* is an *acoustic interferometer* that employs a set of transducers that emit acoustic waves at frequencies above the range of human hearing (higher than 20 kHz). Another set of transducers picks up the reflected acoustic waves, whose wavelength is a fraction of an inch. If anything in the room changes position, the relative phase of the waves, as received by the various acoustic pickups, changes. This data is sent to a microprocessor, which triggers a presence signal. Compare PROXIMITY SENSING.

PRESSURE SENSING

Robotic *pressure sensing* devices detect and measure force, and can in some instances tell where the force is applied.

In a basic pressure sensor, a pressure-sensitive transducer tells a robot when it collides with something. Two metal plates are separated by a layer of nonconductive foam. This forms a capacitor. The capacitor is combined with a coil (inductor). The coil/capacitor circuit sets the frequency of an oscillator. The transducer is coated with plastic to keep the metal from shorting out to anything. If an object hits the sensor, the plate spacing changes. This changes the capacitance, and therefore the oscillator frequency. When the object moves away from the transducer, the foam springs back, and the plates return to their original spacing. This device can be fooled by metallic objects. If a good electrical conductor comes near the transducer, the capacitance might change even if contact is not made.

Conductive foam, rather than dielectric foam, can be placed between the plates, so that the resistance changes with pressure. A direct current is passed through the device. If something bumps the transducer, the current increases because the resistance drops. This transducer will not react to nearby conductive objects unless force is actually applied.

The output of a pressure sensor can be converted to digital data using an *analog-to-digital converter*. This signal can be used by the robot controller. Pressure on a transducer in the front of a robot might cause the machine to back up; pressure on the right side might make the machine turn left.

See also BACK PRESSURE SENSOR, CAPACITIVE PRESSURE SENSOR, CONTACT SENSOR, ELASTOMER, PROXIMITY SENSING, and TACTILE SENSING.

PRINTED CIRCUIT

A printed circuit is a wiring arrangement made of foil on a circuit board. Printed circuits can be mass-produced inexpensively and efficiently. They are compact and reliable. Most electronic devices today are built using printed-circuit technology.

Printed circuits are fabricated by first drawing an *etching pattern*. This is photographed and reproduced on clear plastic. The plastic is placed over a copper-coated glass-epoxy or phenolic board, and the assembly undergoes a photochemical process. The copper dissolves in certain areas, leaving the desired circuit as a pattern of *foil runs*.

The use of printed circuits has vastly enhanced the ease with which electronic equipment can be serviced. Printed circuits allow modular construction, so that an entire board can be replaced in the field and repaired in a fully equipped laboratory.

See also MODULAR CONSTRUCTION.

PROBLEM REDUCTION

Complex problems can be made easier to solve by breaking them down into small steps. This process is called *problem reduction.* It is an important part of research in *artificial intelligence* (AI).

Two common forms

The proof of a mathematical theorem is a good exercise in problem reduction. Another way to develop this skill is to write computer programs in a high-level language.

When breaking a large, difficult problem down into small, easy steps, one can lose sight of the overall picture. Keeping a mental image of the goal, the progress being made, and the obstacles to come is a skill that gets better and better with practice. You cannot sit down and prove profound theorems in mathematics until you have learned to prove some simple things first. The same holds true for smart computers and robots.

Theorem-proving machine

Suppose you build a *theorem-proving machine* (TPM) and assign it a proposition, the proof of which is possible but difficult. Often, a mathematician does not know, when setting out to prove something, if the proposition is true. Thus, the mathematician does know if he or she can solve the problem. In the example shown in the illustration, there are four

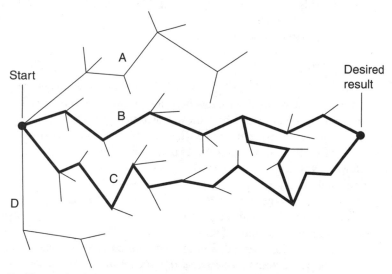

Problem reduction

starting paths: A, B, C, and D. Two of these, B and C, lead to the desired result; the other two do not. But even if TPM starts out along B or C, there are many possible dead ends.

In this example, there is a crossover between paths B and C. One of the sidetracks from path B can lead to the desired result indirectly, by finishing through path C. Also, a sidetrack from path C can take TPM to the proof by moving over to path B. However, these crossings-over can also lead TPM back toward the starting point, and possibly even to dead ends on the way back there.

Dead ends

When TPM runs into a dead end, it can stop, turn around, and backtrack. But how can TPM know that it has come to a dead end? It might keep trying repeatedly to break through the barrier without success. As you know from real-life experience, sometimes persistence can get you over a difficult hurdle, and in other cases all your effort cannot break through the barrier. After you try for a long time to get out of a dead end, you will give up from exasperation and turn back. At what point should TPM give up?

The answer to this quandary lies in the ability of TPM to learn from experience. This is one of the most advanced concepts in AI.

A true TPM that can always solve proofs of true propositions will never, and in fact can never, be constructed. This is because there are statements in any logical system that cannot be proven true or false in a finite number of steps. This was proven by logician Kurt Gödel in 1930, and is called the *incompleteness theorem*.

See also INCOMPLETENESS THEOREM.

PROPRIOCEPTOR

If you close your eyes and move your arms around, you can always tell where your arms are. You know if your arms are raised or whether they are hanging at your sides. You know how much your elbows are bent, the way your wrists are turned, and whether your hands are open or closed. You know which of your fingers are bent and which ones are straight. You know these things because of the nerves in your arms, and the ability of your brain to interpret the signals the nerves send out.

There are advantages in a robot having some of this same sense, so that it can determine, and act according to, its positioning relative to itself. A *proprioceptor* is a system of sensors that allows this.

See also the following definitions: COMPUTER MAP, DIRECTION FINDING, DIRECTION RESOLUTION, DISPLACEMENT TRANSDUCER, DISTANCE MEASUREMENT, DISTANCE RESOLUTION, EDGE DETECTION, EPIPOLAR NAVIGATION, EYE-IN-HAND SYSTEM, GUIDANCE SYSTEM, LANDMARK, LOCAL FEATURE FOCUS, ODOMETRY, PARALLAX, PHOTOELECTRIC PROXIMITY SENSOR, PROXIMITY SENSING, SONAR, and VISION SYSTEM.

PROSODIC FEATURES

In human speech, meaning is conveyed by inflection (tone of voice) as well as by the actual sounds uttered. Perhaps you have heard primitive *speech synthesis* devices with their monotone, emotionless quality. You could understand the words perfectly, but they lacked the changes in pitch, timing, and loudness that give depth to spoken statements. These variations are called *prosodic features.*

To illustrate the importance of prosodic features, consider the sentence, "You will go to the store after midnight." Try emphasizing each word in turn:

- *You* will go to the store after midnight.
- You *will* go to the store after midnight.
- You will *go* to the store after midnight.
- You will go *to* the store after midnight.
- You will go to *the* store after midnight.
- You will go to the *store* after midnight.
- You will go to the store *after* midnight.
- You will go to the store after *midnight.*

Now, instead of making a statement, ask a question, again emphasizing each word in turn. Just replace the period with a question mark. You have 16 different prosodic variations on this one string of words. A few of them are meaningless or silly, but the differences among most of them are striking.

Prosodic variations are important in *speech recognition.* This is because, if you say something one way, you might mean something entirely different than if you utter the same series of words another way. Programming a machine to pick up these subtle differences is one of the greatest challenges facing researchers in artificial intelligence.

See also SPEECH RECOGNITION and SPEECH SYNTHESIS.

PROSTHESIS

A *prosthesis* is an artificial limb or part for the human body. Robotics has made it possible to build electromechanical arms, hands, and legs to replace the limbs of amputees. Artificial organs have also been made. Mechanical legs have been developed to the point where they can let a person walk. Artificial hands can grip; prosthetic arms can throw a ball.

Some internal organs can be replaced, at least for short periods, by machines. Kidney dialysis is one example. An artificial heart is another. Some electronic or electromechanical devices do not completely replace human body parts, but help living organs do what they are supposed to do. An example is a heart pacemaker.

One of the biggest problems with prostheses is that the body sometimes rejects them as foreign objects. The human immune system, which protects against disease, treats the machine as a deadly virus or bacteria, and attempts to destroy it. This puts life-threatening stress on the body. To keep this from happening, doctors sometimes give drugs to suppress the action of the immune system. However, this can make the person more susceptible to diseases such as pneumonia and various viral infections.

Prostheses have not yet been developed that have refined tactile sense. Primitive texture sensing might be developed, but will it ever be as discerning as the real sense of touch? This depends on whether electronic circuits can duplicate the complex impulses that travel through living nerves.

Put a penny and a dime in your pocket. Reach in and, by touch alone, figure out which is which. This is easy; the dime has a ridged edge, but the penny's edge is smooth. This data goes from your fingers to your brain as nerve impulses. Can these impulses be duplicated by electromechanical transducers? Many researchers think so, just as Alexander Graham Bell believed that voice waveforms could be duplicated by electronic devices.

See also BIOMECHANISM and BIOMECHATRONICS.

PROXIMITY SENSING

Proximity sensing is the ability of a robot to tell when it is near an object, or when something is near it. This sense keeps a robot from running into things. It can also be used to measure the distance from a robot to some object.

Basic principle

Most proximity sensors work the same way: the output of a displacement transducer varies with the distance to some object. This can take either of two forms, as shown in the graphs. At left, the sensor output decreases as the distance gets larger. At right, the sensor output rises with increasing distance.

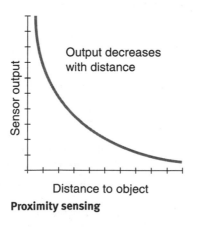
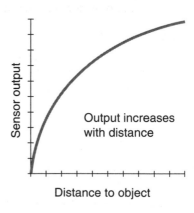

Proximity sensing

In theory, either type of displacement transducer can function in any application, but one type is usually easier to work with, in a given situation, than the other.

Capacitance and inductance

The presence of nearby objects can cause mutual capacitance or mutual inductance effects. These effects can be sensed and the signals transmitted to the robot controller. *Capacitive proximity sensors* operate using electrostatic effects, while *inductive proximity sensors* employ ferromagnetic coupling.

Ladar

An infrared (IR) or visible laser beam can be bounced off anything that will reflect or scatter the energy. The return-signal delay can be measured, and the distance to the object determined by the robot controller. This is called *ladar* (short for *laser detection and ranging*).

Ladar will not work for objects that do not reflect IR or visible energy. A white painted wall will reflect such energy well; the same wall painted dull black will not. Ladar works better at relatively long distances than at short distances, over which *sonar* or *interferometry* provide superior results.

Radar and sonar

Proximity sensing can be done using *radar* or *sonar*. Radar works with ultra-high-frequency (UHF) or microwave radio signals. Sonar uses acoustic waves. Pulses are transmitted and picked up after they reflect from objects. The delay time is measured, and the results sent to the robot controller. The principle is basically like that of a laser-ranging proximity sensor.

Radar will not work for objects that do not reflect UHF or microwave energy. Metallic objects reflect this energy well; salt water is fair; and trees and houses are poor. Radar, like ladar, works better at long distances than close up. Sonar can function well at small distances, because the speed of sound in much slower than the speed of electromagnetic (EM) waves in free space.

For further information

Related definitions, besides those already mentioned here, include ACOUSTIC PROXIMITY SENSOR, ARTIFICIAL STIMULUS, CAPACITIVE PROXIMITY SENSOR, COMPUTER MAP, DISTANCE MEASUREMENT, GUIDANCE SYSTEMS, INDUCTIVE PROXIMITY SENSOR, LADAR, PARALLAX, PHOTOELECTRIC PROXIMITY SENSOR, PRESENCE SENSING, RADAR, RANGE SENSING AND PLOTTING, SONAR, and VISION SYSTEM.

Q

QUADRUPED ROBOT

Historically, people have been fond of the idea of building a robot in the human image. Such a machine has two legs. In practice, a two-legged, or bipedal, robot is difficult to design. It tends to have a poor sense of balance; it falls over easily. The sense of balance, which humans take for granted, is difficult to build into a machine. (Specialized two-wheel robots have been designed that incorporate a sense of balance, but they are sophisticated and costly.)

To guarantee stability, a robot that uses legs for locomotion must always have at least three feet in contact with the surface. A four-legged machine, called a *quadruped robot,* can pick up one leg at a time while walking, and remain stable. The only problem occurs when the three surface-bound legs lie on or near a common line, as shown on the left side of the illustration. Under these conditions, a four-legged object can topple.

In the best quadruped design, the four feet reach the ground at points that are not near a common line, as shown on the right side of the illustration. Then, when one foot is lifted for propulsion, the other three are

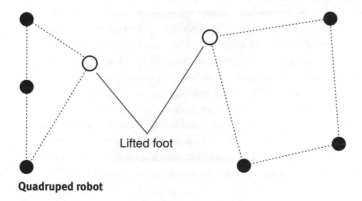

Lifted foot

Quadruped robot

on the surface at the vertices of a well-defined triangle. In these examples, the solid dots represent feet on the ground; the open circle represents the foot that is lifted at the moment.

Many engineers believe that six legs is optimal for robots designed to propel with legs rather than by rolling on wheels or a track drive. Six-legged robots can lift one or two legs at a time while walking and remain stable. The more legs a robot has, the better its stability; but there is a practical limit. The movements of robot legs must be coordinated properly for a machine to propel itself without wasting motion and energy. This becomes increasingly difficult as the number of legs increases.

See also **INSECT ROBOT** and **ROBOT LEG.**

QUADTREE

A *quadtree* is a scheme in which a two-dimensional (2-D) rectangular *occupancy grid* can be divided into smaller and smaller subelements, as necessary to define a function to a desired level of resolution. The illustration shows a simple example. In this case, the robot's work environment (or *world space*) is shown by the largest square. This square is divided into four square subelements. The upper left subelement is in turn split into four square sub-subelements (or sub^2-elements); the lower right

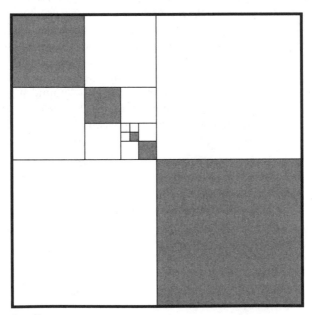

Quadtree

sub^2-element is divided into four sub^3-elements; the lower right sub^3-element is divided into four sub^4-elements; the upper left sub^4-element is divided into four square sub^5-elements. This process can continue until the limit of resolution, or the required level of precision, is reached.

If the world space is not square or rectangular in shape, the situation becomes more complicated. However, a *bit map* of square elements can approximate a 2-D world space of any shape, provided the elements are small enough.

If a robot's world space is three-dimensional (3-D), it can be divided into cubes or rectangular prisms (blocks). Each block can be divided into eight subblocks. This process can be repeated in the same manner as the 2D quadtree. The result is called an octree.

See also COMPUTER MAP and OCCUPANCY GRID.

QUALITATIVE NAVIGATION

See TOPOLOGICAL NAVIGATION.

QUALITY ASSURANCE AND CONTROL (QA/QC)

In factory work, robots can perform repetitive tasks more accurately, and faster, than human workers. Robotization has improved the quality, as well as increased the quantity, of production in many industries.

Doing it better

An important, but often overlooked, aspect of quality assurance and control lies in the production process itself. One way to ensure perfect quality is to do a perfect job of manufacturing. Robots are ideal for this. Not all robots work faster than humans, but robots are almost always more consistent and reliable. When the manufacturing process is improved, fewer faulty units come off an assembly line. This makes *quality assurance and control (QA/QC)* comparatively easy.

Some QA/QC engineers say that, in an ideal world, their jobs would not be necessary. Flawed materials should be thrown away before they are put into anything. Assembly robots should do perfect work. This philosophy has been stated by Japanese QA/QC engineer Hajime Karatsu: "Do such good work that QA/QC checkers are not necessary." This is a theoretical ideal, of course; manufacturing processes are not, and will never be, perfect. There will always be errors in assembly, or defective components that get into production units. Thus, there will always be a need for at least one QA/QC person to keep bad units from getting to buyers.

Inspectors

Robots can sometimes work as QA/QC engineers. However, they can do this only for simple inspections, because QA/QC work often requires that the inspector have a keen sense of judgment.

One simple QA/QC job is checking bottles for height as they move along an assembly line. A laser/robot combination can pick out bottles that are not the right height. The principle is shown in the illustration. If a bottle is too short, both laser beams reach the photodetectors. If a bottle is too tall, neither laser beam reaches the photodetectors. In either of these situations, a robot arm/gripper picks the faulty bottle off the line and discards it. Only when a bottle is within a very narrow range of heights (the acceptable range) will one laser reach its photodetector while the other laser is blocked. Then the bottle is allowed to pass.

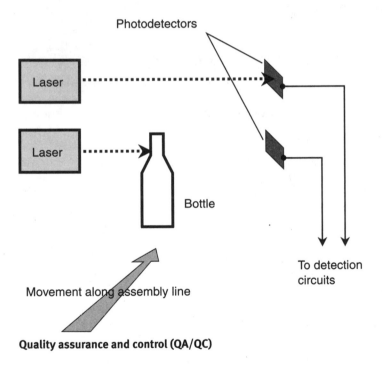

Quality assurance and control (QA/QC)

Robotic QA/QC processes are becoming more complex and sophisticated with the advancement of *artificial intelligence* (AI). But some QA/QC decisions involve intuition. This sense is common in people, but engineers question whether any machine can be programmed to have it.

Some computers can learn from their mistakes and make informed decisions based on large amounts of data, but the ability to "follow a hunch" appears to be a quality unique to humans.

QUANTITATIVE NAVIGATION

See **METRIC PATH PLANNING.**

R

RADAR

Electromagnetic waves at radio frequencies (RF) are reflected from metallic objects. The term *radar* is a contraction of the full technical description, *radio detection and ranging*. Radar can be used by robots as a navigation aid, and also for speed measurement.

A radar system intended for ranging and direction measurement consists of a transmitter, a directional antenna, a receiver, and a position indicator. The transmitter produces intense pulses of RF *microwaves*. These waves strike objects. Some things (such as cars and trucks) reflect radar waves better than others (such as wood). The reflected signals, or *echoes,* are picked up by the antenna. The farther away the reflecting object, the longer the time before the echo is received. The transmitting antenna is rotated so that the radar sees in all directions.

As the radar antenna rotates, echoes are received from various directions. In a robot, these echoes are processed by a microcomputer that gives the machine a sense of its position relative to the work environment. Radar can be used by robotic aircraft and spacecraft.

A special form of radar, called *Doppler radar,* is used to measure the speed of an approaching or retreating target, or the speed of a robot with respect to a barrier. This type of radar operates by means of the Doppler effect, as shown in the illustration on page 254. This is how police radar measures the speed of an oncoming vehicle. Compare LADAR and SONAR.

See also COMPUTER MAP, DIRECTION RESOLUTION, DISTANCE RESOLUTION, and RANGE SENSING AND PLOTTING.

RADIANT HEAT DETECTOR

See PRESENCE SENSING.

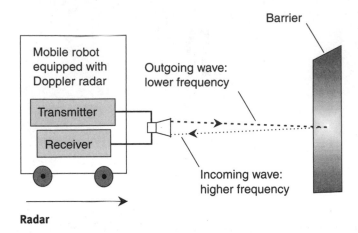

Radar

RADIO-FREQUENCY INTERFERENCE (RFI)

Radio-frequency interference (RFI) is a phenomenon in which electronic devices upset each other's operation. In recent years this problem has been getting worse because consumer electronic devices are proliferating, and they have become increasingly susceptible to RFI.

Much RFI results from inferior equipment design. To some extent, faulty installation methods also contribute to the problem. Computers produce wideband radio-frequency (RF) energy that is radiated if the computer is not well shielded. Computers can malfunction because of strong RF fields, such as those from a nearby broadcast transmitter. This can, and often does, happen when the broadcast transmitter is working perfectly. In these cases, and also in cases involving cellular telephones, citizens' band (CB) radios, and amateur ("ham") radios, the transmitting equipment is almost never at fault; the problem is almost always improper or ineffective shielding of the computer system.

RFI is often picked up on power and interconnecting cables. There are methods of bypassing or choking the RF on these cables, preventing it from getting into the computer, but the bypass or choke must not interfere with the transmission of data through cables. For advice, consult the dealer or manufacturer of the computer.

Power lines can cause RFI. Such interference is almost always caused by arcing. A malfunctioning transformer, or a bad street light, or a salt-encrusted insulator can all be responsible. Often, help can be obtained by calling the utility company.

A *transient suppressor,* also called a *surge suppressor,* in the power cord is essential for reliable operation of a personal computer or robot controller operating from utility lines. A *line filter,* consisting of capacitors between each side of the power line and ground, can help prevent RF from getting into a computer via the utility lines.

As computers become more portable and more common, RFI problems can be expected to worsen unless manufacturers pay stricter attention to electromagnetic shielding. As computers are increasingly used as robot controllers, potential problems multiply. An errant robot can create a hazard and cause accidents. The danger is greatest with medical or life-support devices.

RANGE

Range is the distance, as measured along a straight line in a specific direction in three-dimensional (3-D) space, between a robot and an object or barrier in the work environment. In the case of a sensor, the range is the maximum radial distance over which the device can be expected to work properly.

In mathematics and logic, the term *range* refers to the set of objects (usually numbers) onto which objects in the *domain* of a mathematical *function* are mapped.

See also FIELD OF VIEW (FOV), RANGE OF FUNCTION, and RANGE SENSING AND PLOTTING.

RANGE IMAGE

See DEPTH MAP.

RANGE OF FUNCTION

The *range* of a mathematical *function* is the set of things (usually numbers) onto which objects in the *domain* are mapped. Every x in the domain of a function f is mapped onto exactly one value y. There might be, and often are, y values that do not have anything mapped onto them by function f. These points are outside the range of f.

Suppose you are given the function $f(x) = +x^{1/2}$ (that is, the positive square root of x) for $x > 0$. The graph of this function is shown in the illustration. This function always maps x onto a positive real number y. No matter what value you pick for x in the domain in this example, $+x^{1/2}$ is positive.

Computers work extensively with functions, both analog and digital. Functions are important in robotic navigation, location, and measurement systems.

See also DOMAIN OF FUNCTION and FUNCTION.

RANGE PLOTTING

Range plotting is a process in which a graph is generated depicting the distance (range) to objects, as a function of the direction in two or three dimensions.

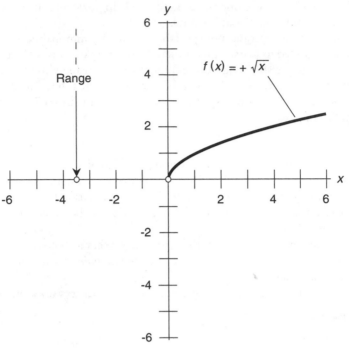

Range of function

To do *one-dimensional (1-D) range plotting,* a signal is sent out, and the robot measures the time it takes for the echo to come back. This signal can be an acoustic wave, in which case the device is *sonar.* Or it can be a radio wave; this is *radar.* If it is visible light in the form of a laser beam, it is *ladar.*

Two-dimensional (2-D) range plotting involves mapping the distances to various objects, as a function of their direction in a defined plane. One method is shown in the illustration. The robot is at the center of the plot, in a room containing three desks (rectangles) and two floor lamps (circles). The range is measured every 10° of azimuth around a complete circle, resulting in the set of points shown. A better plot would be obtained if the range were plotted every 5°, every 2°, or even every 1° or less. But no matter how detailed the *direction resolution,* the 2-D range plot can show things in only one plane, such as the floor level or some horizontal plane above the floor.

Three-dimensional (3-D) range plotting requires the use of *spherical coordinates.* The distance must be measured for a large number of directions at all orientations. A 3-D range plot in a room such as that depicted in the

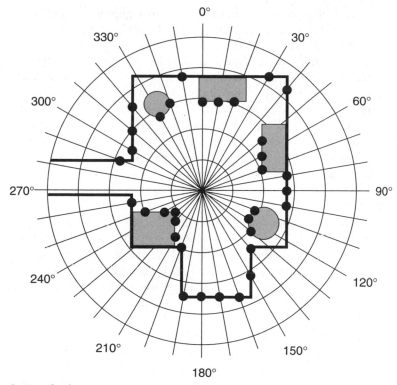

Range plotting

illustration would show ceiling fixtures, objects on the floor, objects on top of the desks, and other details not visible with a 2-D range plot.

See also COMPUTER MAP, DEPTH MAP, DIRECTION RESOLUTION, DISTANCE RESOLUTION, LADAR, RADAR, and SONAR.

RANGING

See DISTANCE MEASUREMENT.

REACTIVE PARADIGM

The *reactive paradigm* is an approach to robot programming in which all actions are the direct results of sensor output. No advance planning is involved. This approach arose because of limitations inherent in the *hierarchical paradigm,* which relies on rigid adherence to a specific plan in order to reach a goal. The reactive paradigm became popular around 1990, and its use was favored through the early 1990s.

In the most sophisticated robot systems, there are three basic functions, known as *plan/sense/act.* The reactive paradigm simplifies this to *sense/act.*

A robot that operates in this fashion is analogous to a human or animal that exhibits reflex actions when certain stimuli occur.

The principal asset of the reactive paradigm is high speed. Just as human or animal reflexes occur faster than behaviors that involve conscious thought (deliberation), robots using the reactive paradigm can respond to changes in their environment almost instantly. However, there are drawbacks to this approach. The simple sense/act approach can sometimes result in cycling back and forth between two conditions, without making any progress toward the intended goal. This can be considered the robotic equivalent of human disorientation or panic. Compare HIERARCHICAL PARADIGM and HYBRID DELIBERATIVE/REACTIVE PARADIGM.

REAL TIME

In communications or data processing, operation done "live" is called *real-time* operation. The term applies especially to computers. Real-time data exchange allows a computer and the operator to converse.

Real-time operation is convenient for storing and verifying data within a short time. This is the case, for example, when making airline reservations, checking a credit card, or making a bank transaction. However, real-time operation is not always necessary. It is a waste of expensive computer time to write a long program at an active terminal. Long programs are best written off-line, tested in real time (on-line), and debugged off-line.

In a fleet of *insect robots* all under the control of a single computer, real-time operation can be obtained for all the robots simultaneously. One method of achieving this is *time sharing*. The controller pays attention to each robot for a small increment of time, constantly rotating among the robots at a high rate of speed. Compare TIME SHIFTING.

RECTANGULAR COORDINATE GEOMETRY

See CARTESIAN COORDINATE GEOMETRY.

RECURSION

Recursion is a logical process in which one or more tasks are set aside while the main argument is being made. Recursion is common in computer programs, where it can take the form of *nested loops*. Recursion is also useful in proving mathematical and legal propositions. It is a powerful tool in *artificial intelligence* (AI).

Keep final goal in mind

Recursion can be intricate, and is one of the most advanced forms of human reasoning. For recursion to work, the overall direction of progress is toward the final goal. Sidetracking might seem to have nothing to do with the intended result, but in recursion, there is always a reason for it. All the subarguments must eventually be brought out and put to some use in the main argument.

Computers are ideally suited for recursive arguments. The subarguments can be done and the results put into memory. Humans get confused when there are too many sidetracks; not so with computers. They will do precisely what they are programmed to do, and they do not get distracted, no matter how many sidetracks there are.

In a complicated recursive argument, the sidetracks can be backed up one on top of another, like airplanes in a holding pattern waiting to land at a large airport. The subarguments are held in *pushdown stacks,* or *first-in/last-out* memory registers. The sidetracked results are pulled out of the stacks when needed. The illustration shows a recursive argument with several pushdown stacks.

Hangups

If a computer uses recursive logic and gets sidetracked too much, it might lose sight of the final objective, or go around and around in logical circles. When this happens in a computer program, it is called an *endless loop* or an *infinite loop.* This is makes it impossible to solve any problem.

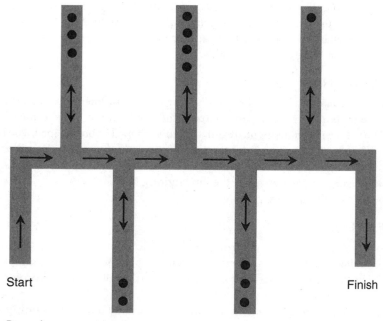

Start

Finish

Recursion

There is another logical trap into which humans can easily fall when making a recursive argument. This is to "prove" something by unknowingly making the assumption that it is already true. Computers, properly programmed, do not make this mistake.

REDUCTIONISM

Reductionism is the hypothesis that all human thought can be duplicated by machines. Can all human thought and emotion ultimately be reduced to logical ones and zeros? A reductionist would say yes.

The human brain is far more complicated than any computer yet devised, but the brain is made of a finite number of individual cells. For any finite number, no matter how large, there exists a larger number. If a brain has, say, the equivalent of 10^{25} *logic gates,* then there can be, at least in theory, a computer chip with 10^{25} logic gates. The reductionist argues that all human mental activity is nothing more than the sum total of many gates working in many ways. Even though the number might be gigantic, it is nevertheless finite.

Reductionism is of interest to *artificial intelligence* (AI) researchers. If the reductionist hypothesis proves true, then computers might be made into living entities. Some researchers are enthusiastic about this, and others are concerned about the possible negative implications. Science fiction authors have exploited this theme; perhaps the earliest example was a play called *Rossum's Universal Robots,* written in 1920 by Karel Capek. In this play, which the author intended as a satire, robots become alive and take over the world.

REGULAR GRID

A *regular grid* is a method of dividing up a two-dimensional (2-D) work environment into square or rectangular regions. In three-dimensional (3-D) environments, the regions are cube-shaped or box-shaped. The basis for the rectangular grid is the *Cartesian coordinate system,* also called the rectangular coordinate system. This is the familiar *xy* plane or *xyz* space of analytic geometry (see the illustration).

See also QUADTREE.

REINITIALIZATION

Sometimes a robot controller will operate improperly because of stray voltages. When this happens, the microcomputer malfunctions or becomes inoperative. *Reinitialization* consists of setting all of the microcomputer lines to low or zero.

Most microcomputers are automatically reinitialized every time power is removed and reapplied. Not all microcomputers have this feature, how-

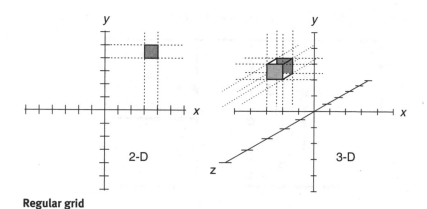

Regular grid

ever; a specific procedure must be followed to reinitialize such devices.
See **CONTROLLER**.

RELATIONAL GRAPH

A *relational graph* is a representation of a robotic work environment as
points, called *nodes,* and lines connecting those points, called *edges.* A
relational graph is generated on the basis of a *computer map.*

Consider a simple floor plan, such as that shown in the illustration. A
basic relational graph can be generated by locating the center points of all
the rooms, and the center points of all the doorways, and defining each
such point as a node. If there is a turn in the hallway, a point midway
from the protruding corner to the opposite wall, subtending an angle of
135° with either wall at the corner, is defined as a node. These nodes are
then connected with straight-line edges.

Relational graphs provide a means of robot navigation in environments
that do not change geometrically, and in which no new obstructions are
placed. However, this type of graph does not generally represent the most
efficient navigation method, and it can be inadequate for large robots or
for fleets of robots in a limited space.

See also **COMPUTER MAP, GATEWAY, LANDMARK,** and **TOPOLOGICAL PATH PLANNING.**

RELIABILITY

Reliability is an expression of how well, and for how long, machines keep
working. It is the proportion of units that still work after they have been
used for a certain length of time.

Suppose that 1,000,000 units are placed in operation on January 1,
2010. If 920,000 units are operating properly on January 1, 2011, then the
reliability is 0.92, or 92 percent, per year. On January 1, 2012, you can

Floor plan with nodes and edges

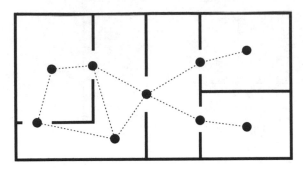

Nodes and edges only

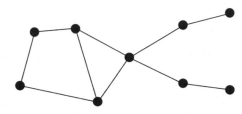

Relational graph

expect that 920,000 × 0.92 = 846,400 units will be working. The number of working units declines according to the reliability factor, year after year.

The better the reliability, the flatter is the decay curve in a graph of working units versus time. This is shown in the illustration. The terms "excellent," "good," "fair," and "poor" are relative, and depend on many factors. A perfect reliability curve (100 percent) is always a horizontal line on such a graph.

Reliability is a function of design, as well as of the quality of the parts and the precision of the manufacturing process. Even if a machine is well made, and the components are of good quality, failure is more likely with poor design than with good design. Reliability can be optimized by *quality assurance and control.*

See also QUALITY ASSURANCE AND CONTROL (QA/AC).

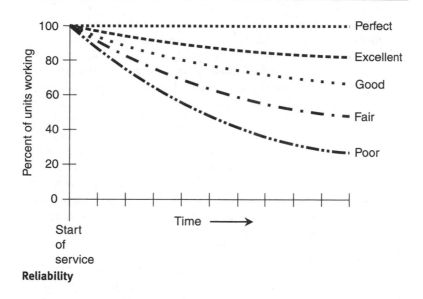

REMOTE CONTROL

Robots can be operated from a distance by human beings. Computers can also be controlled from places far removed from the machines themselves. This is done by means of *remote control*.

A simple example of a remote-control system is the control box for a television (TV) set. Another example is a transmitter used to fly a model airplane. The TV control employs infrared (IR) radiation to carry the data. The model airplane gets its commands via radio signals. In this sense, both the TV set and the model airplane are robots.

Remote control can be done by wire, cable, or optical fiber links. Undersea robots have been operated in this way. A person sits at a terminal in the comfort of a boat or submarine bubble and operates the robot, watching a screen that shows what the robot "sees." This is a crude form of *telepresence*. The range of remote control is limited when wires or optical fibers are used. It is impractical to have a cable longer than a few kilometers. A special problem exists for long-distance undersea remote control. Radio waves at conventional radio frequencies cannot penetrate the oceans, but extremely long cables present mechanical problems.

When the control station and the robot are very far away from each other, even radio, IR, or visible-light signals take a long time to cover the distance. A remotely controlled robot on the moon is about 1.3 light-seconds away. It is 2.6 s from the time a command is sent to a robot on the moon until the operator sees the results of the command.

One of the most dramatic examples of radio remote control is the *trasmission of* commands to space probes as they fly through the solar system. In these cases, the separation distance is of the order of millions of kilometers. As the probe *Voyager* passed Neptune, and a command was sent to the probe, the results were not observed for hours. Remote control of this type is a special challenge.

There is an absolute limit to the practical distance that can exist between a remotely controlled robot and its operator. There is (as yet) no known way to transmit data faster than the speed of electromagnetic (EM) energy in free space. Related definitions include AUTONOMOUS ROBOT, FLIGHT TELEROBOTIC SERVICER, FLYING EYEBALL, GUIDANCE SYSTEM, INSECT ROBOT, LASER DATA TRANSMISSION, MICROWAVE DATA TRANSMISSION, REMOTELY OPERATED VEHICLE, REMOTE MANIPULATOR, ROBOTIC SPACE TRAVEL, SECURITY ROBOT, SELSYN, SYNCHRO, TELEOPERATION, and TELEPRESENCE.

REPULSIVE RADIAL FIELD

See POTENTIAL FIELD.

RESOLUTION

Resolution is the ability of a robotic *vision system* to distinguish between things that are close together. Within objects, resolution is the extent to which the system can bring out details about the object. It is a precise measure of image quality. It is sometimes called *definition*.

In a robotic vision system, the resolution is the "sharpness" of the image. Poor resolution can be the result of poor focus, too few pixels in the image, or a signal bandwidth that is not wide enough. The illustration shows two objects that are far away and close together, as they might appear to a robot vision system having four different levels of resolution.

When an analog image is converted into digital form, the *sampling resolution* is the number of different digital levels that are possible. This number is generally some power of 2. An analog signal has infinitely many different levels; it can vary over a continuous range. The higher the sampling resolution, the more accurate is the digital representation of the signal.

In *position sensing*, and also in *range sensing and plotting*, the terms *direction resolution* and *distance resolution* refer to the ability of a robot sensor to differentiate between two objects that are separated by a small angle, or that are almost the same distance away. The term *spatial resolution* refers to the smallest linear displacement over which a robot can define its work environment and correct errors in its motion.

See also DIRECTION RESOLUTION, DISTANCE RESOLUTION, PIXEL, SPATIAL RESOLUTION, and VISION SYSTEM.

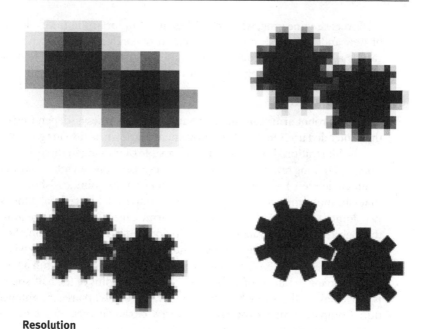

Resolution

REVERSE ENGINEERING

It is possible to build a machine that does the same things as some other machine, but using a different design. When this is done with computers, it is called *cloning*. In general, complex or sophisticated devices or systems have more equivalent designs than simple devices or systems. *Reverse engineering* is a process by which a device or system is copied functionally, but not literally.

Reverse engineering raises legal issues. If you can duplicate the things a patented machine will do, but use a new and different approach that you thought of independently, you do not, in most cases, infringe on the patent of the original machine. If you invent something like a smart robot and then get it patented, you cannot normally get a patent for what it does. For example, you cannot design a bicycle-waxing robot and then expect to get a patent that will keep anyone else from legally building and selling a robot that can wax bicycles.

But suppose someone reverse-engineers a patented product by dismantling it and then rebuilding it almost, but not quite, the same way. This person does not invent a new design. The work is used in slightly, but not significantly, altered form, and then a claim is made that the resulting product is "new." This constitutes patent infringement.

Reverse engineering, when done legally, is important in the evolution of new and improved robotic systems. In research and development, it can be a valuable technique in hardware design, programming, and the development of operating systems for robot controllers.

REVOLUTE GEOMETRY

Industrial robot arms can move in various different ways, depending on their intended use. One mode of movement is known as *revolute geometry.*

The illustration shows a robot arm capable of moving in three dimensions (3-D) using revolute geometry. The entire assembly can rotate through a full circle (360°) at the base. There is an elevation joint, or "shoulder," that can move the arm through 90°, from horizontal to vertical. One or two joints in the middle of the robot arm, called "elbows," can move through 180°, from a straight position to doubled back. As an option, there can be a "wrist" that freely rotates either clockwise or counterclockwise.

A well-designed revolute robot arm can reach any point within a half-sphere having the shape of an inverted bowl. The radius of the half-sphere is the length of the arm when its shoulder and elbow(s) are straightened out. Compare CARTESIAN COORDINATE GEOMETRY, CYLINDRICAL COORDINATE GEOMETRY, POLAR COORDINATE GEOMETRY, and SPHERICAL COORDINATE GEOMETRY.

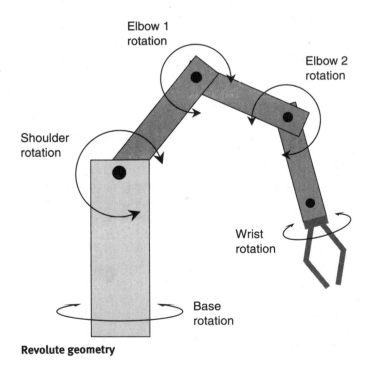

Revolute geometry

ROBOT ARM

There are numerous ways in which a *robot arm* can be designed. Different configurations are used for different purposes. Some robots, especially industrial robots, are nothing more than sophisticated robot arms. Robot arms are sometimes called *manipulators,* although technically this term applies to the arm and its *end effector,* if any.

A robot arm can be categorized according to its geometry. Two-dimensional (2-D) designs have *work envelopes* limited to a section of a flat plane. Most robot arms can work within a region of three-dimensional (3-D) space.

Some robot arms resemble human arms. The joints in these machines are given names like "shoulder," "elbow," and "wrist." Some robot arms, however, are so very different from human arms that these names do not make sense. An arm that employs *revolute geometry* is similar to a human arm, but an arm that uses *Cartesian coordinate geometry* is far different. For further information, see ARTICULATED GEOMETRY, CARTESIAN COORDINATE GEOMETRY, CYLINDRICAL COORDINATE GEOMETRY, POLAR COORDINATE GEOMETRY, REVOLUTE GEOMETRY, SPHERICAL COORDINATE GEOMETRY, and WORK ENVELOPE.

ROBOT CLASSIFICATION

In the late twentieth century, the *Japan Industrial Robot Association (JIRA)* classified robots from simple manipulators to advanced systems incorporating *artificial intelligence* (AI). From low-end to high-end, the JIRA *robot classification* scheme proceeds as follows:

1. *Manually operated manipulators:* Machines that must be directly operated by a human.
2. *Sequential manipulators:* Devices that perform a series of tasks in the same sequence every time they are actuated. A good example is a telephone answering machine.
3. *Programmable manipulators:* These include the simpler types of industrial robots familiar to most people.
4. *Numerically controlled robots:* Examples are servo robots.
5. *Sensate robots:* Robots incorporating sensors of any type, such as back pressure, proximity, pressure, tactile, or wrist force.
6. *Adaptive robots:* Robots that adjust the way they work to compensate for changes in their environment.
7. *Smart robots:* Robots with high-end controllers that can be considered to possess AI.
8. *Intelligent mechatronic systems:* Computers that control a fleet of robots or robotic devices.

Some researchers and engineers add another category: *intelligent bio-mechatronic systems*. These include devices such as *cyborgs* and certain *prostheses*.

See also **ROBOT GENERATIONS**.

ROBOT GENERATIONS

Engineers and scientists have analyzed the evolution of robots, marking progress according to *robot generations*.

First generation

A *first-generation robot* is a simple mechanical arm. These machines have the ability to make precise motions at high speed, many times, for a long time. Such robots find widespread industrial use today.

First-generation robots can work in groups, such as in an *automated integrated manufacturing system (AIMS)*, if their actions are synchronized. The operation of these machines must be constantly supervised, because if they get out of alignment and are allowed to keep working, the result can be a series of bad production units.

Second generation

A *second-generation robot* has rudimentary machine intelligence. Such a robot is equipped with sensors that tell it things about the outside world. These devices include *pressure sensors, proximity sensors, tactile sensors, radar, sonar, ladar,* and *vision systems*. A controller processes the data from these sensors and adjusts the operation of the robot accordingly. These devices came into common use around 1980.

Second-generation robots can stay synchronized with each other, without having to be overseen constantly by a human operator. Of course, periodic checking is needed with any machine, because things can always go wrong; the more complex the system, the more ways it can malfunction.

Third generation

The concept of a *third-generation robot* encompasses two major avenues of evolving *smart robot* technology: the *autonomous robot* and the *insect robot*.

An autonomous robot can work on its own. It contains a controller, and it can do things largely without supervision, either by an outside computer or by a human being. A good example of this type of third-generation robot is the *personal robot* about which some people dream.

There are some situations in which autonomous robots do not perform efficiently. In these cases, a fleet of simple insect robots, all under the control of one central computer, can be used. These machines work like ants in an anthill, or like bees in a hive. While the individual machines lack *artificial intelligence* (AI), the group as a whole is intelligent.

Fourth generation and beyond

Any robot of a sort yet to be seriously put into operation is a *fourth-generation robot*. Examples of these might be robots that reproduce and evolve, or that incorporate biological as well as mechanical components. Past that, we might say that a *fifth-generation robot* is something no one has yet designed or conceived.

The table summarizes robot generations, their times of development, and their capabilities.

Robot generations: comparison of features

Generation	Time First Used	Capabilities
First	Before 1980	Mechanical
		Stationary
		Good precision
		High speed
		Physical ruggedness
		Use of servomechanisms
		No external sensors
		No artificial intelligence
Second	1980–1990	Tactile sensors
		Vision systems
		Position sensors
		Pressure sensors
		Microcomputer control
		Programmable
Third	Mid-1990s and after	Mobile
		Autonomous
		Insectlike
		Artificial intelligence
		Speech recognition
		Speech synthesis
		Navigation systems
		Teleoperated

Robot generations: comparison of features (Cont.)

Generation	Time First Used	Capabilities
Fourth	Future	Design not yet begun
		Able to reproduce?
		Able to evolve?
		Artificially alive?
		As smart as a human?
		True sense of humor?
Fifth	?	Not yet discussed
		Capabilities unknown

ROBOT GRIPPER

A *robot gripper* is a specialized *end effector* that can take either of two general forms: handlike, and non-handlike. These two main schemes arise from different engineering philosophies.

Some researchers say that the human hand is an advanced device, having evolved by natural selection. Therefore, they say, robotics engineers should imitate human hands when designing and building robot grippers. Other roboticists argue that specialized grippers should be used, because robots ordinarily must do only a few specific tasks. Human hands are used for many things, but such versatility might be unnecessary, and even detrimental, in a robot made for a single task.

For more information about robot grippers and related subjects, see ACTIVE CHORD MECHANISM, BACK PRESSURE SENSOR, END EFFECTOR, EYE-IN-HAND SYSTEM, FINE MOTION PLANNING, GRASPING PLANNING, LADLE GRIPPER, JAW, JOINT FORCE SENSING, PITCH, PRESSURE SENSING, PROPRIOCEPTOR, PROSTHESIS, ROLL, SERVOMECHANISM, TACTILE SENSING, TEXTURE SENSING, TWO-PINCHER GRIPPER, VACUUM CUP GRIPPER, WRIST FORCE SENSOR, and YAW.

ROBOT HEARING

See BINAURAL ROBOT HEARING, DYNAMIC TRANSDUCER, ELECTROSTATIC TRANSDUCER, PIEZOELECTRIC TRANSDUCER.

ROBOTIC SHIP

A modern passenger jet can be, and to a large extent is, flown by a computer. It has been said that such an aircraft could take off from New York, fly to Sydney, and land without a single human on board. Such an aircraft is, in effect, a robot. In a similar way, ocean-going vessels can be controlled by computers.

A robotic ship might be designed for combat, and built solely for the purpose of winning battles at sea. With no humans on board, there would be no risk to human lives. The ship would require no facilities for people, such as sleeping quarters, food service, and medical service. The only necessity would be to protect the robot controller from damage. Imagine being the captain of a destroyer, and going up against another destroyer that had no humans on board! Such an enemy would have no fear of death and, therefore, would be extremely dangerous.

Robots are playing an increasing role in military applications, but most experts doubt that passenger transports will ever be fully roboticized.

See also ROBOTIC SPACE TRAVEL.

ROBOTIC SPACE TRAVEL

The U.S. space program climaxed when *Apollo 11* landed on the Moon and, for the first time, a creature from Earth walked on another world. Some people think the visitor from Earth could just as well have been, and should have been, a robot.

Some types of spacecraft have been remotely controlled for decades. Communications satellites use radio commands to adjust their circuits and change their orbits. Space probes, such as the *Voyager* that photographed Uranus and Neptune in the late 1980s, are controlled by radio. Satellites and space probes are crude robots.

Space probes work like other hostile-environment machines. Robots are used inside nuclear reactors, in dangerous mines, and in the deep sea. All such robots operate by means of *remote control*. The remote-control systems are getting more and more sophisticated as technology improves.

Almost like being there

Some people say that robots should be used to explore outer space, while people stay safely back on Earth and work the robots by means of *teleoperation* or *telepresence*. A human operator can wear a special control suit and have a robot mimic all movements. Teleoperation is the simple remote-control operation of a robot. Telepresence involves remote control with continuous feedback that gives the operator a sense of being in the robot's place.

Some roboticists believe that with technology called *virtual reality*, it is possible to duplicate the feeling of being in a remote location, to such an extent that the robot operator can imagine that he or she is really there. Stereoscopic vision systems, binaural hearing, and a crude sense of touch can be duplicated. Imagine stepping into a gossamer-thin suit, walking into a chamber, and existing, in effect, on the Moon or Mars, free of danger from extreme temperatures or deadly radiation.

The main problem

If robots are used in space travel, with the intention of having the machines replace astronauts, then the distance between the robot and its operator cannot be very great. The reason is that the control signals cannot travel faster than 299,792 km/s (186,282 mi/s), the speed of light in free space.

The Moon is approximately 400,000 km, or 1.3 light-seconds, from Earth. If a robot, not Neil Armstrong, had stepped onto the Moon on that summer day in 1969, its operator would have had to deal with a delay of 2.6 s between command and response. It would take each command 1.3 s to get to the Moon, and each response 1.3 s to get back to Earth. True telepresence is impossible with a delay like that. Experts say that the maximum delay for true telepresence is 0.1 s. The distance between the robot and its controller thus cannot be more than 0.5, or 1/20, of a light-second. That is about 15,000 km or 9300 mi—slightly more than the diameter of Earth.

A possible scenario

Suppose that astronauts are in orbit around a planet whose environment is too hostile to allow an in-person visit. Then a robot can be sent down. An example of such a planet is Venus, whose crushing surface pressures would kill an astronaut in any pressure suit possible with current technology. It would be easy to sustain an orbit of less than 9300 mi above Venus, however, so telepresence would be feasible. The operator could sit in a spacecraft in orbit above the planet, and get the feeling of walking around on the surface.

See also TELEOPERATION and TELEPRESENCE.

ROBOT LEG

A *robot leg* is an appendage similar to a robot arm, but intended to support and propel a mobile robot rather than manipulate objects. *Legged locomotion* has advantages when the terrain in a robot's world space is irregular or rough. Legs can also allow robots to jump, sit down, and kick objects. However, wheel or track drives are usually preferable in work environments with smooth, comparatively level surfaces.

Humans have dreamed of building machines in their own image. In reality, humanoid robots almost always are built for amusement. When robots have legs, stability is a concern. A robot can fall over if it must stand on one or two legs, or if all its legs are lined up.

Legged robots usually have four or six legs. The legs can be independently maneuverable, or they can move in groups. Robots with more than six legs have not often been conceived.

See also BIPED ROBOT, INSECT ROBOT, QUADRUPED ROBOT, TRACK-DRIVE LOCOMOTION, TRI-STAR WHEEL LOCOMOTION, and WHEEL-DRIVE LOCOMOTION.

ROBOT VISION

See **VISION SYSTEM.**

ROLL

Roll is one of three types of motion that a robotic *end effector* can make. It is a rotational form of motion, unlike *pitch* and *yaw*, which are back-and-forth (or up-and-down) movements.

Extend your arm out straight, and point at something with your index finger. Twist your wrist. Your index finger keeps pointing in the same direction, but it rotates along with your wrist. If your index finger were the head of a screwdriver, it would be able to turn a screw. This is an example of roll. Compare **PITCH** and **YAW.**

RULE-BASED SYSTEM

See **EXPERT SYSTEM.**

S

SATELLITE DATA TRANSMISSION

Satellite data transmission is a form of *microwave data transmission,* but the repeaters are in space, not on the ground. Signals are sent up to the satellite, received, and retransmitted on another frequency at the same time. The ground-to-satellite data are called the *uplink*; the satellite-to-ground data are the *downlink*. Satellite data transmission can be used in the remote control of robots over great distances, and in outer space.

Many satellites are in geostationary orbits, at fixed spots 36,000 km above Earth's equator. When such a satellite is used, the total path length is always at least twice this. The smallest possible delay is therefore approximately ¼ s. High-speed, two-way data communication is impossible with a path delay that long, as is realistic *telepresence*. However, *teleoperation* (simple remote control) of robots is possible.

The illustration shows a system that uses two geostationary satellites

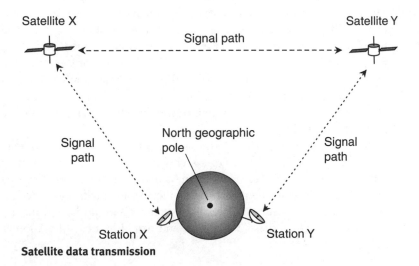

Satellite data transmission

to achieve data communication between two locations that are nearly at opposite points on the surface of Earth. Station X can be considered the location of the control operator, and station Y the location of a remotely controlled robot.

One of the biggest challenges facing researchers in *artificial intelligence* (AI) is how to link computers that are separated by vast distances. There is no way to overcome the fact that the speed of light is slow on a large scale, and when considered in terms of the time required for a computer to execute a clock cycle.

See also MICROWAVE DATA TRANSMISSION.

SCALING

Scaling is a principle familiar to structural engineers and physicists. As an object is made larger to an equal extent in all linear dimensions, its structural integrity diminishes.

When things get larger but stay in the same relative proportions, mechanical strength increases as to the square (second power) of linear dimension—height, width, or depth. However, mass increases according to the cube (third power) of the linear dimension. The illustration shows how this works with cubes. The mass, and thus the weight in a constant gravitational field, goes up faster than the linear dimension or cross-sectional area increases. Eventually, if an object gets large enough, it becomes physically unstable or mechanically unworkable.

Consider a theoretical solid cube of variable size but perfectly homogeneous matter. In the illustration, the smaller cube has height = 1 unit, width = 1 unit, and depth = 1 unit. The larger cube is double this size in each linear dimension: height = 2 units, width = 2 units, and depth = 2 units. The base (or cross-sectional) area of the smaller cube is 1 unit squared (1×1); the volume of the smaller cube is 1 unit cubed $(1 \times 1 \times 1)$. The base (or cross-sectional) area of the larger cube is 4 units squared (2×2); the volume is 8 units cubed $(2 \times 2 \times 2)$. If the cubes are made of the same homogeneous material, doubling the linear dimension also doubles the weight per unit surface area at the base. As the cube keeps getting larger, it will eventually fall through or sink into the surface, or collapse under its own weight.

Imagine the situation with a humanoid robot. If its height suddenly increases by a factor of 10, its cross-sectional structural area increases by a factor of $10^2 = 100$. However, its mass becomes $10^3 = 1000$ times as great. That is the equivalent of a 10-fold increase in gravitational acceleration. A robot constructed of ordinary materials would have difficulty maneuvering under these conditions, and would be unstable. Another

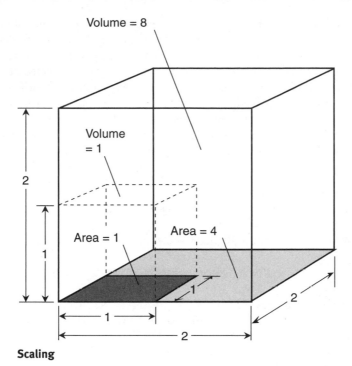

Volume = 8

Volume = 1

Area = 1

Area = 4

Scaling

10-fold increase in linear dimension would result in physical collapse. This is why giant robots are unwieldy and impractical, while small ones are comparatively hardy and durable.

SECURITY ROBOT

The term *security robot* refers to any robot that assists in the protection of people or property, particularly against crime. Security robots have existed for decades. A simple version is the electronic garage-door opener. If you lock yourself out of your house, you can gain entry through the garage door if you have the control box in your possession. (So can anyone with a control box that operates on the same frequency and has the same signaling code.)

Mid-level security robots include intrusion alarm systems, electronic door/gate openers that employ digital codes, and various surveillance systems. These devices can discourage unauthorized people from entering a property. If someone does gain entry, a mid-level system can detect the presence of an intruder, usually by means of ultrasound, microwaves, or lasers, and notify police through a telephone or wireless link.

A hypothetical high-end security system of the future consists of one or more mobile robots that resemble servants some of the time and attack dogs at other times. The system will minimize the opportunity or desirability of intrusion. If an unauthorized person enters the protected property, the security robots will drive the offender away, or detain the offender until police arrive. Robots of this type have been depicted in movies. Because of these movies, some people believe that such machines will someday become commonplace. However, there are numerous problems with this scheme. Here, posed in the form of questions, are some examples of the challenges facing designers of the ultimate robotic security system.

- Can such robots be fast enough, and have good enough vision, to chase down an intruder or win a fight with a human being who is in good physical condition?
- Can such robots be designed to detect any intruder at any time?
- Can such robots be tamper-proof?
- Can such robots be designed to withstand an assault with practically any weapon?
- If the above questions can all be answered "Yes," will the cost of a system of this level ever be affordable to the average family or small business?
- Will property owners be able to trust their security robots to work all the time?
- What if the robot malfunctions and thinks the owner is an intruder?
- Can a machine lawfully use deadly force?
- What will be the consequences if a security robot injures or kills an intruder?

See also POLICE ROBOT and SENTRY ROBOT.

SEEING-EYE ROBOT

Mobile smart robots have been suggested as possible replacements for seeing-eye dogs. An advanced machine can help visually impaired people navigate their surroundings.

A so-called *seeing-eye robot* must have a vision system with excellent sensitivity and resolution. The robot must have *artificial intelligence* (AI) at least equivalent to the intelligence of a dog. The machine must be able to negotiate all types of terrain, doing such diverse things as crossing a street, passing through a crowded room, or climbing stairs.

The Japanese, with their enthusiasm for robots that resemble living creatures, have designed various seeing-eye robots. They are approximately the same size as live dogs. Most roll on wheels or track drives.

See also PERSONAL ROBOT.

SELSYN

A *selsyn* is an indicating device that shows the direction in which an object is pointing. It consists of a position sensor and a transmitting unit at the location of the movable device, and a receiving unit and indicator located in a convenient place. A common application of the selsyn is as a direction indicator for a rotatable sensor, as shown in the illustration.

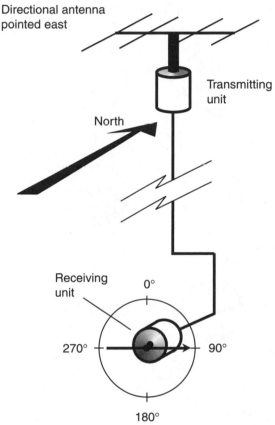

Selsyn

In a selsyn, the indicator usually rotates through the same number of angular degrees as the moving device. A selsyn for azimuth bearings normally rotates 360°; a selsyn for elevation bearings turns through 90°. Compare SERVOMECHANISM, STEPPER MOTOR, and SYNCHRO.

SEMANTIC NETWORK

A *semantic network* is a reasoning scheme that can be used in *artificial intelligence* (AI). In a semantic network, the objects, locations, actions, and tasks are called *nodes*. The nodes are interconnected by *relations*. This breaks reasoning down in a manner similar to the way sentences are diagrammed in grammar analysis. The main difference is that a semantic network is not limited to any single sentence; it can build on itself indefinitely, so that it represents more and more complex scenarios.

An example of a semantic network is shown in the illustration. The nodes are circles, and the relations are lines connecting the circles. The situation can be inferred. Additions can be made. (Use your imagination.)

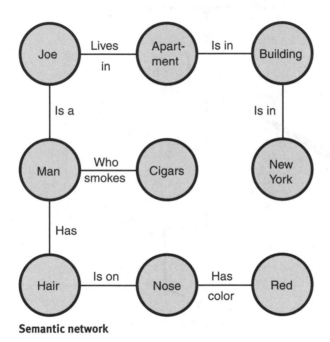

Semantic network

Some researchers believe that semantic networks are more versatile than another common reasoning device, known as *expert systems*. Compare EXPERT SYSTEM.

See also IF/THEN/ELSE.

SENSE/PLAN/ACT

See HIERARCHICAL PARADIGM, HYBRID DELIBERATIVE/REACTIVE PARADIGM, and REACTIVE PARADIGM.

SENSOR COMPETITION

In some robotic systems, more than one sensor is used to detect a single *percept*, or stimulus in the environment. *Sensor competition* is the use of two or more redundant sensors to minimize the number of *false negatives* and *false positives.*

Whenever a single sensor is used to detect a phenomenon or occurrence in the environment, there is a chance for error. If the output of the sensor is a simple "yes/no" (logic 1 or logic 0), the output will occasionally be 1 when it should be 0 (false positive) or vice versa (false negative). If a sensor detects a range of values such as visible-light intensity, the measurement is always subject to some error.

Suppose two binary-output (1 or 0) sensors are used to detect or measure the same phenomenon. The output of the combination can be considered 1 if and only if both sensors output 1; the output of the combination can be considered 0 if and only if both sensors output 0. Usually the two sensors will agree, but occasionally they will not. In cases where the two sensors disagree, the robot controller can instruct the sensors to take another sampling. In the case of analog sensors, such as those used to measure visible-light intensity, the outputs can be averaged to get a more accurate reading than either sensor produces alone.

Numerous competing sensors can be used to obtain much greater accuracy than is possible with a single sensor. In general, the larger the number of competing sensors, the less frequent will be the errors in a binary digital system, and the smaller will be the error in an analog system. There are various ways in which the sensor outputs can be combined to obtain a result of the desired accuracy, while maintaining reasonable system speed.

SENSOR FUSION

The term *sensor fusion* refers to the use of two or more different types of sensors simultaneously to analyze an object. Examples of characteristics that can be measured include mass (or weight), volume, shape, light reflectivity, light transmittivity, color, temperature, and texture.

Sensor fusion is used by smart robots to identify objects. The robot controller can store a large database of objects and their unique characteristics. When an object is encountered, the sensors provide input and compare the characteristics of the object with the information in the database. Compare SENSOR COMPETITION.

See also BACK PRESSURE SENSOR, COLOR SENSING, EYE-IN-HAND SYSTEM, FEEDBACK, JOINT-FORCE SENSING, OBJECT RECOGNITION, TACTILE SENSING, TEMPERATURE SENSING, TEXTURE SENSING, and WRIST-FORCE SENSING.

SENTRY ROBOT

A *sentry robot* is a specialized type of *security robot* that alerts people to abnormal conditions. Such a robot can be designed to detect smoke, fires, burglars, or flooding. A sentry robot might detect abnormal temperature, barometric pressure, wind speed, humidity, or air pollution.

In industry, sentry robots can alert personnel to the fact that something is wrong. The robot might not specifically pinpoint and identify the problem, but it can let people know that a system is malfunctioning. A fire, for example, generates smoke and/or infrared (IR), either or both of which can be detected by a roving sentry.

A high-end sentry robot might include features such as:

- Air-pressure sensing
- Autonomy
- Beacon navigation
- Computer map(s) of the environment
- Guidance systems
- Homing devices
- Intrusion detection
- Ladar
- Mobility
- Position sensing
- Radar
- Wireless links to controller and central station
- Smoke detection
- Sonar
- Speech recognition
- Tactile sensing
- Temperature sensing
- Vision systems

SERIAL DATA TRANSMISSION

See DATA CONVERSION.

SERVOED SYSTEM

See CLOSED-LOOP SYSTEM.

SERVOMECHANISM

A *servomechanism* is a specialized feedback-control device. Servomechanisms are used to control mechanical things such as motors, steering mechanisms, and robots.

Servomechanisms are used extensively in robotics. A robot controller can tell a servomechanism to move in certain ways that depend on the inputs from sensors. Multiple servomechanisms, when interconnected and controlled by a sophisticated computer, can do complex tasks such as cook a meal. A set of servomechanisms, including associated circuits and hardware, and intended for a specific task, constitutes a *servo system*. Servo systems do precise, often repetitive, mechanical chores.

A computer can control a servo system made up of many servomechanisms. For example, an unmanned robotic warplane (also known as a *drone*) can be programmed to take off, fly a mission, return, and land. Servo systems can be programmed to do assembly-line work and other tasks that involve repetitive movement, precision, and endurance.

A *servo robot* is a robot whose movement is programmed into a computer. The robot follows the instructions given by the program, and carries out precise motions on that basis. Servo robots can be categorized according to the way they move. In *continuous-path motion,* the robot mechanism can stop anywhere along its path. In *point-to-point motion,* it can stop only at specific points in its path. Servo robots can be easily programmed and reprogrammed. This might be done by exchanging diskettes, by manual data entry, or by more exotic methods such as a *teach box.* Compare SELSYN, STEPPER MOTOR, and SYNCHRO.

See also CLOSED-LOOP SYSTEM, CONTINUOUS-PATH MOTION, OPEN-LOOP SYSTEM, PERSONAL ROBOT, POINT-TO-POINT MOTION, and TEACH BOX.

SHAPE-SHIFTING ROBOT

See POLYMORPHIC ROBOT.

SHARED CONTROL

Shared control, also called *continuous assistance,* is a form of robotic remote control in a system that employs *teleoperation.* The operator oversees the execution of a complex task such as repairing a satellite on a Space Shuttle mission. The human operator can delegate some portions of the task to the robot, but supervision must be maintained at all times. If necessary, the operator can intervene and take control of (assist) the robot.

Shared control has assets in certain situations, especially critical missions. The human operator constantly monitors the progress of the machine. The system can contend with sudden, unforeseen changes in the work environment.

Shared control has limitations. It is difficult for a single operator to oversee the operation of more than one robot at a time. *Latency,* or the time lag caused by signal propagation delays, makes two-way teleoperation difficult if the operator and the robot are separated by a great distance. Shared control is impractical, for example, in the teleoperation of a robot on the other side of the solar system. Still another problem is that large signal bandwidth is required during those periods when the human operator must take direct control of the robot. In scenarios such as these, *control trading* is generally superior to shared control. Compare CONTROL TRADING.

See also TELEOPERATION.

SIDE LIGHTING

In a robotic vision system, the term *side lighting* refers to illumination of objects in the work environment using a light source located such that the scene is lit up from one side, or from the top or the bottom. The light from the source scatters from the surfaces of the objects under observation before reaching the sensors. In addition, the robot sees significant shadow effect in its work environment.

Side lighting is used in situations where the surface details of observed objects are of interest or significance. This scheme lends a sense of depth to a scene because of the shadows cast by objects. Irregularities in a surface show up especially well when the illumination strikes the surface at a sharp angle. (A good example is the illumination of the craters in the twilight zone on the Moon, as seen through a telescope, when the Moon is in its first-quarter or last-quarter phase.) Side lighting does not work well in situations involving translucent or semitransparent objects, if their internal structure must be analyzed. *Back lighting* works best in these cases. Compare BACK LIGHTING and FRONT LIGHTING.

SIGNAL GENERATOR

See GENERATOR.

SIMPLE-MOTION PROGRAMMING

As machines become smarter, the programming gets more sophisticated. No machine has yet been built that has intelligence anywhere near that of a human being. Some researchers think that true *artificial intelligence* (AI), at a level near that of the human brain, will never be achieved.

The programming of robots can be divided into levels, starting with the least sophisticated and progressing to the theoretical level of true AI. The drawing shows a four-level scheme. Level 1, the lowest level, is *simple-motion programming*. Robots at this level are designed to perform basic, often repetitive actions, such as actuating a motor or lifting an object. Compare ARTIFICIAL INTELLIGENCE, COMPLEX-MOTION PROGRAMMING, and TASK-LEVEL PROGRAMMING.

SIMULATION

Simulation is the use of computers to mimic real-life situations. Some simulators involve teaching of skills for the operation of machinery. Other simulators are programs that predict (or try to predict) events in the real world.

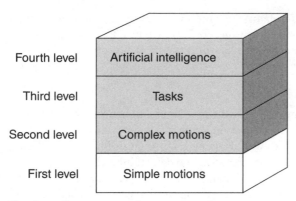

Fourth level	Artificial intelligence
Third level	Tasks
Second level	Complex motions
First level	Simple motions

Simple-motion programming

An *interactive simulator* resembles a high-end video game. In fact, computerized video games nowadays are more sophisticated than some simulations. There is usually a video monitor, a set of controls, and a set of indicators. There may also be audio devices and motion-imitation machines. The controls depend on the scenario.

Suppose you get into a simulator intended to mimic the experience of a driver in the Indy 500 auto race. The controls include an accelerator, brakes, and a steering wheel. There is a speedometer and a tachometer. There are speakers that emit noises similar to those a real driver would hear. The seat vibrates and/or rocks back and forth. A high-resolution display screen renders a perspective-enhanced view of the virtual road, virtual cars, and the virtual surroundings as they whiz by. Interactive simulation is often used as a teaching/training aid for complex skills, such as flying an aircraft. This technique is especially useful in the military, for training in a wide variety of skills.

An *event simulator* is a computer program that imitates, or models, the behavior of a system. For example, you might want to start a business. How well will it operate? Will you go bankrupt? Will you make a million dollars in your first year? The event simulator, if it is sophisticated enough and if it is given enough data, can help provide answers to questions like these.

One of the most important event simulators is the hurricane forecasting model employed by the National Hurricane Center in Miami, Florida. As Hurricane Andrew approached in August 1992, the computers predicted the most likely places for landfall. Andrew took an unusual, east-to-west path. Hurricanes often curve northward before they strike land, but the Hurricane Center model predicted that Andrew would keep going due west until it had passed over the Florida peninsula. The event simulator in this case proved accurate.

As event simulators become more advanced, they will increasingly incorporate *artificial intelligence* (AI) to draw their conclusions, but there will always be an element of uncertainty that limits the effectiveness of event simulators.

SMART HOME

Imagine having all your mundane household chores done without your having to think about them! A home computer could control a fleet of personal robots, which would take care of cooking, dish washing, the laundry, yard maintenance, snow removal, and other things. This is the ultimate form of computerized home: having a "central nervous system" run by a computer. In the building trade, a computerized home is called a *smart home*.

Technology and ethics

The key to a smart home lies in the technologies of robotics and *artificial intelligence* (AI). As these become more available to the average consumer, we can expect to see, for example, robotic laundry stations. Robots will be available to make our beds, do our dishes, vacuum our carpets, shovel snow from our driveways, and clean our windows.

There are two main types of mobile robots that might roam the home of the future: *autonomous robots* and *insect robots*. There are advantages and disadvantages to either design. In addition to these, some appliances will themselves be robots, such as dishwashers and laundry machines.

Some people question whether computerized, robotized homes are worth developing. Some people will prefer to spend their hard-earned money in other ways, such as buying vacations or new properties. There are also ethical concerns. Should some people strive for total home automation, when a large segment of society cannot afford a home at all?

Let us suppose, for the moment, that we solve the ethical problem, and that everyone has a home and some money to spare. Further imagine that the cost of technology keeps going down, while it keeps getting more and more sophisticated. What might the future hold?

Protection from fire

When people and property must be protected from fire, *smoke detection* is a simple and effective measure. Smoke detectors are inexpensive, and can operate on flashlight batteries. You should have one or more of these devices in your home now.

In a computerized home of the future, a smoke alarm could alert a robot. Robots are ideal for fire-fighting because they can do things that are too dangerous for humans to do. The challenge will be to program the robots to have judgment comparable to that of human firefighters.

When and if household robots become commonplace, one of their duties will be to ensure the safety of the human occupants. This will include escorting people from the house if it catches fire, and then putting out the fire and/or calling the fire department. It might also involve performing simple first-aid tasks.

Security

Computers and robots can be of immense help around the house when it comes to prevention of burglaries.

Security robots have been around for decades. A simple version is the electronic garage-door opener. More advanced systems include intrusion alarm systems and electronic door/gate openers. High-end devices make it difficult for unauthorized people to enter a property.

A *sentry robot* can alert a homeowner to abnormal conditions. It might detect fire, burglars, or water. A sentry might detect abnormal temperature, barometric pressure, wind speed, humidity, or air pollution.

Food service

Robots can prepare and serve food. So far, the major applications have been in repetitive chores, such as placing measured portions on plates, cafeteria style, to serve a large number of people. However, robots can be adapted to food service in common households.

Personal robots, when they are programmed to prepare or serve food, require more autonomy than robots in large-volume food service. You might insert a disk into a home robot that tells it to prepare a meal of meat, vegetables, and beverages, and perhaps also dessert and coffee. The robot would ask you questions, such as:

- How many people will be here for supper tonight?
- Which type of meat would you like?
- Which type of vegetable?
- How would you like the potatoes done? Or would you rather have rice?
- What beverages would you like?

When all the answers are received, the robot executes the tasks necessary to prepare the meal. The robot might serve you as you wait at the table, and then clean up the table when you are done eating. It might do the dishes, too.

Yard work

Riding mowers and riding snow blowers will be easy for robots to use. The robot need only sit on the chair, ride the machine around, and operate the handlebar/pedal controls. Alternatively, lawn mowers or snow blowers can be robotic devices themselves, designed with the applicable task in mind.

The main challenge, once a lawn-mowing or snow-blowing robot has begun its work, is for it to do its work everywhere that it should, but nowhere that it should not. You don't want the lawn mower in your garden, and there is no point in blowing snow from your lawn. Current-carrying wires might be buried around the perimeter of your yard, and along the edges of the driveway and walkways, establishing the boundaries within which the robot must work.

Inside the work area, edge detection can be used to follow the line between mown and unmown grass, or between cleared and uncleared pavement. This line is easily seen because of differences in brightness and/or color. Alternatively, a computer map can be used, and the robot can sweep along controlled and programmed strips with mathematical precision.

The idle homeowner

If robots can do all our housework, what will be left for us to do? Won't people get bored sailing, hiking, working out, and otherwise spending time that used to be devoted to maintaining our property?

Although robots and computers can do work for us, we do not have to employ them. There will always be times when people prefer to do household chores themselves. Many people enjoy doing their own gardening. Perhaps the greatest challenge in home automation will be to decide what tasks are best left to the homeowners.

A major hangup with home computerization is the matter of trust. Most people have trouble enough entrusting computers with simple tasks such as storing data. Some people will never be comfortable leaving a computer or robotic system entirely in charge of the house.

See also AUTONOMOUS ROBOT, FIRE-PROTECTION ROBOT, INSECT ROBOT, PERSONAL ROBOT, SECURITY ROBOT, SENTRY ROBOT, and SMOKE DETECTION.

SMART ROBOT

See AUTONOMOUS ROBOT.

SMOKE DETECTION

When people and property must be protected from fire, *smoke detection* is a simple and effective measure. Smoke detectors are inexpensive, and can operate on flashlight batteries.

Smoke changes the characteristics of the air. It is accompanied by changes in the relative amounts of gases. Fire burns away oxygen and produces other gases, especially carbon dioxide. The smoke itself consists of solid particles.

Air has a property called the *dielectric constant*. This is a measure of how well the atmosphere can hold an electric charge. Air also has an *ionization potential*; this is the energy required to strip electrons from the atoms. Many things can affect these properties of air. Common factors are humidity, pressure, smoke, and changes in the relative concentrations of gases.

A smoke detector can work by sensing a change in the dielectric constant, and/or the ionization potential, of the air. Two electrically charged plates are spaced a fixed distance apart (see the illustration). If the properties of the air change, the plates gain or lose some of their electric charge. This causes momentary currents that can actuate alarms or robotic systems.

See also FIRE-PROTECTION ROBOT and SENTRY ROBOT.

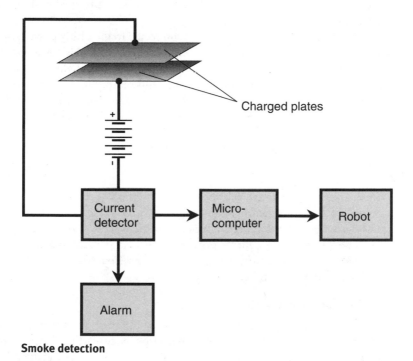

Charged plates

Smoke detection

SOCIETY

See INSECT ROBOT.

SOLAR POWER

Electric power can be obtained directly from sunlight by means of *photo-voltaic cells*. Most photovoltaic cells, also called *solar cells*, produce about

15 mW of power for each square centimeter of surface area exposed to bright sunlight. Solar cells produce direct current (DC), which is used by most electronic systems. Solar power is ideal for use in mobile robots that function outdoors, especially in environments that get plenty of sunshine.

The *stand-alone solar power system* is most suitable for mobile robots. It uses banks of rechargeable batteries, such as the lead–acid type, to store electric energy as it is supplied by photovoltaics during hours of bright sunshine. The energy is released by the batteries at night or in gloomy daytime weather. This system is independent of the electric utility. The main limitation of the stand-alone solar power system for use in mobile robots is the fact that the solar batteries (combinations of solar cells in series and/or parallel) must have significant surface area exposed to the sun in order to generate enough power to operate robotic propulsion motors. This can pose a design problem.

An *interactive solar power system* is connected to the utility power grid. This type of system does not normally use storage batteries. Any excess energy is sold to utility companies during times of daylight and minimum usage. Energy is bought from the utility at night, during gloomy daytime weather, or during times of heavy usage. This scheme can be used with fixed robots, or with computers intended to control fleets of mobile robots.

See also ELECTROCHEMICAL POWER and POWER SUPPLY.

SONAR

Sonar is a medium-range and short-range method of distance measurement. The term is an acronym that stands for *sound detection and ranging*. The basic principle is simple: bounce acoustic waves off of objects, and measure the time it takes for the echoes to return. In practice, sonar systems can be made so sophisticated that they rival *vision systems* for getting pictures of the environment.

Audible versus ultrasonic

Sonar can make use of audible sound waves, but there are advantages to using ultrasound instead. Ultrasound has a frequency too high for humans to hear, ranging from about 20 kHz to more than 100 kHz. (One kilohertz, kHz, is 1000 cycles per second.)

An obvious advantage of ultrasound in robotics is that the acoustic-wave bursts are not heard by people working around the robot. These waves, if audible, can be annoying. Another advantage of ultrasound over audible sound is the fact that a system using ultrasound is less likely to be fooled by people talking, machinery operating, and other common noises. At frequencies higher than the range of human hearing, acoustic disturbances do not occur as often, or with as much intensity, as they do within the hearing range.

A simple sonar

The simplest sonar scheme is shown in the block diagram. An ultrasonic pulse generator sends bursts of alternating current (AC) to a transducer. This converts the currents into ultrasound, which is sent out in a beam. This beam is reflected from objects in the environment, and returns to a second transducer, which converts the ultrasound back into pulses of AC. These pulses are delayed with respect to those that were sent out. The length of the delay is measured, and the data fed to a microcomputer that determines the distance to the object in question.

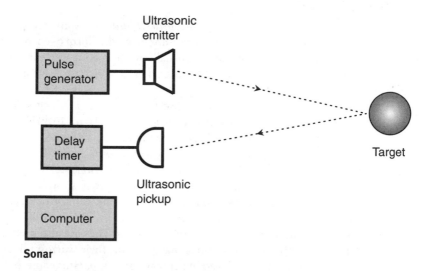

Sonar

This system cannot provide a detailed picture of the environment unless it is refined, and a computer is incorporated to analyze the pulses coming in. Sonar beams tend to be rather broad; acoustic waves are difficult to focus. This limits the *image resolution* obtainable with sonar. Another problem with this simple system is that it can be fooled if the echo delay is equal to, or longer than, the time between individual pulses.

Refinements

Researchers know that high-end sonar systems can rival vision systems as a means of mapping the environment, because bats—whose "vision" is actually sonar—can navigate as well as if they had keen eyesight.

What makes bats so adept at using sonar? For one thing, they have a brain. It follows that *artificial intelligence* (AI) must be an important part of any advanced robotic sonar system. The computer must analyze the incoming pulses in terms of their phase, the distortion at the leading and trailing edges, and whether or not the returned echoes are *bogeys* (illusions or false echoes).

For good image resolution, the sonar beam must be as narrow as possible, and it must be swept around in two or three dimensions. With optimum *direction resolution* and *distance resolution,* a sonar can make a computer map of a robot's work environment. Compare LADAR and RADAR.

See also COMPUTER MAP, DIRECTION RESOLUTION, DISTANCE RESOLUTION, and RANGE SENSING AND PLOTTING.

SOUND TRANSDUCER

A *sound transducer,* also called an *acoustic transducer,* is an electronic component that converts acoustic waves into some other form of energy, or vice versa. The other form of energy is usually an alternating-current (AC) electrical signal. The waveforms of the acoustical and electrical signals are identical, or nearly so.

Acoustic transducers are designed for various frequency ranges. The human hearing spectrum extends from about 20 Hz to 20 kHz, but acoustic energy can have frequencies lower than 20 Hz or higher than 20 kHz. Energy at frequencies below 20 Hz is called *infrasound;* if the frequency is above 20 kHz, it is *ultrasound.* In acoustic wireless devices, ultrasound is generally used, because the wavelength is short and the necessary transducers can be small. Also, ultrasound cannot be heard, and therefore it does not distract or annoy people.

Sound transducers are used in security systems. They are also used in robotics to help mobile machines navigate in their surroundings. Acoustic transducers are employed in depth-finding apparatus commonly found on boats.

See also DYNAMIC TRANSDUCER, ELECTROSTATIC TRANSDUCER, and PIEZOELECTRIC TRANSDUCER.

SPATIAL RESOLUTION

Spatial resolution is a quantitative measure of the detail with which a robot can define its work environment. It can be expressed in meters, centimeters, millimeters, or micrometers (units of 10^{-6} m). In some precision robots, it may be expressed in nanometers (units of 10^{-9} m). This measure can refer to either of two quantities:

- The smallest linear distance between two points that the robot can differentiate

- The edge dimension of the smallest cubical parcel of space the robot can define

In general, the smaller the spatial-resolution number, the greater is the accuracy with which the robot can position its *end effector*(s) or move to a specific location, and the smaller the error that can be sensed and corrected. The spatial resolution of a robotic system depends on the resolution of the controller. As the resolution increases, so does the required amount of memory and processing power. In servomechanisms, the spatial resolution depends on the smallest displacement the device can detect.

See also DISTANCE RESOLUTION, DIRECTION RESOLUTION, and RESOLUTION.

SPEECH RECOGNITION

The human voice consists of audio-frequency (AF) energy, with components ranging from about 100 Hz to several kilohertz (kHz). (A frequency of 1 Hz is one cycle per second; 1 kHz = 1000 Hz.) This has been known ever since Alexander Graham Bell sent the first voice signals over electric wires.

As computer-controlled robots evolve, people naturally want to control them just by talking to them. *Speech recognition,* also called *voice recognition,* makes this possible. The illustration is a block diagram of a simple speech-recognition system.

Components of speech

Suppose you speak into a microphone that is connected to an oscilloscope, and see the jumble of waves on the screen. How can any computer be

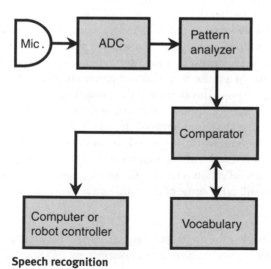

Speech recognition

programmed to make sense out of that? The answer lies in the fact that, whatever you say, it is comprised of only a few dozen basic sounds called *phonemes*. These phonemes can be identified by computer programs.

In communications, a voice can be transmitted if the bandwidth is restricted to the range from 300 to 3000 Hz. Certain phonemes, such as "ssss," contain energy at frequencies of several kilohertz, but all the information in a voice, including the emotional content, can be conveyed if the audio passband is cut off at 3000 Hz. This is the typical voice frequency response in a two-way radio.

Most of the acoustic energy in a human voice occurs within three defined frequency ranges, called *formants*. The first formant is at less than 1000 Hz. The second formant ranges from approximately 1600 to 2000 Hz. The third formant ranges from approximately 2600 to 3000 Hz. Between the formants there are spectral gaps, or ranges of frequencies at which little or no sound occurs. The formants, and the gaps between them, stay in the same frequency ranges no matter what is said. The fine details of the voice print determine not only the words, but all the emotions, insinuations, and other aspects of speech. Any change in "tone of voice" shows up in a voice print. Therefore, in theory, it is possible to build a machine that can recognize and analyze speech as well as any human being.

A/D Conversion

The *passband,* or range of audio frequencies transmitted in a circuit, can be reduced greatly if you are willing to give up some of the emotional content of the voice, in favor of efficient information transfer. *Analog-to-digital conversion* accomplishes this. An *analog-to-digital converter (ADC)* changes the continuously variable, or analog, voice signal into a series of digital pulses. This is a little like the process in which a photograph is converted to a grid of dots for printing in the newspaper. There are several different characteristics of a pulse train that can be varied. These include the pulse amplitude, the pulse duration, and the pulse frequency.

A digital signal can carry a human voice within a passband less than 200 Hz wide. That is less than one-tenth of the passband of the analog signal. The narrower the bandwidth, in general, the more of the emotional content is sacrificed. Emotional content is conveyed by *inflection,* or variation in voice tone. When inflection is lost, a voice signal resembles a monotone. However, it can still carry some of the subtle meanings and feelings.

Word analysis

For a computer to decipher the digital voice signal, it must have a vocabulary of words or syllables, and some means of comparing this knowledge base with the incoming audio signals. This system has two parts: a *memory,* in which various speech patterns are stored; and a *comparator,* which compares these stored patterns with the data coming in. For each syllable or word, the circuit checks through its vocabulary until a match is found. This is done very quickly, so the delay is not noticeable. The size of the computer's vocabulary is related directly to its memory capacity. An advanced speech-recognition system requires a large amount of memory.

The output of the comparator must be processed in some way, so that the machine knows the difference between words or syllables that sound alike. Examples are "two/too," "way/weigh," and "not/knot." For this to be possible, the *context* and *syntax* must be examined. There must also be some way for the computer to tell whether a group of syllables constitutes one word, two words, three words, or more. The more complicated the voice input, the greater is the chance for confusion. Even the most advanced speech-recognition system makes mistakes, just as people sometimes misinterpret what you say. Such errors will become less frequent as computer memory capacity and operating speed increase.

Insinuations and emotions

The ADC in a speech-recognition system removes some of the inflections from a voice. In the extreme, all of the tonal changes are lost, and the voice is reduced to "audible text." For most robot-control purposes, this is adequate. If a system could be 100 percent reliable in just getting each word right, speech-recognition engineers would be very pleased. However, when accuracy does approach 100 percent, there is increasing interest in getting some of the subtler meanings across, too. Consider the sentence, "You will go to the store after midnight," and say it with the emphasis on each word in turn (eight different ways). The meaning changes dramatically depending on the *prosodic features* of your voice: which word or words you emphasize. Tone is important for another reason, too: a sentence might be a statement or a question. Thus, "You will go to the store after *midnight?*" represents something completely different from "You *will* go to the store after midnight!" Even if all the tones are the same, the meaning can vary depending on how quickly something is said. Even the timing of breaths can make a difference.

For further information

Speech recognition is a rapidly advancing technology. The best source of up-to-date information is a good college library. Ask the librarian for reference books, and for articles in engineering journals, concerning the most recent developments. A search on the phrases "speech recognition" and "voice recognition" can be conducted on the Web using Google (www.google.com) or a similar search engine. Related entries include BANDWIDTH, CONTEXT, DATA CONVERSION, DIGITAL SIGNAL PROCESSING, MESSAGE PASSING, OPTICAL CHARACTER RECOGNITION, PROSODIC FEATURES, SOUND TRANSDUCER, SPEECH SYNTHESIS, and SYNTAX.

SPEECH SYNTHESIS

Speech synthesis, also called *voice synthesis,* is the electronic generation of sounds that mimic the human voice. These sounds can be generated from digital text or from printed documents. Speech can also be generated by high-level computers that have *artificial intelligence* (AI), in the form of responses to stimuli or input from humans or other machines.

What is a voice?

All audible sounds consist of combinations of alternating-current (AC) waves within the frequency range from 20 Hz to 20 kHz. (A frequency of 1 Hz is one cycle per second; 1 kHz = 1000 Hz.) These take the form of vibrations in air molecules. The patterns of vibration can be duplicated as electric currents.

A frequency band of 300 to 3000 Hz is wide enough to convey all the information, and also all of the emotional content, in any person's voice. Therefore, speech synthesizers only need to make sounds within the range from 300 to 3000 Hz. The challenge is to produce waves at exactly the right frequencies, at the right times, and in the right phase combinations. The modulation must also be correct, so the intended meaning is conveyed. In the human voice, the volume and frequency rise and fall in subtle and precise ways. The slightest change in modulation can make a tremendous difference in the meaning of what is said. You can tell, even over the telephone, whether the speaker is anxious, angry, or relaxed. A request sounds different than a command. A question sounds different than a declarative statement, even if the words are the same.

Tone of voice

In the English language there are 40 elementary sounds, known as *phonemes.* In some languages there are more phonemes than in English; some languages have fewer phonemes. The exact sound of a phoneme can vary, depending on what comes before and after it. These variations are called *allophones.* There are 128 allophones in English. These can be strung together in myriad ways.

The inflection, or "tone of voice," is another variable in speech; it depends on whether the speaker is angry, sad, scared, happy, or indifferent. These depend not only on the actual feelings of the speaker, but on age, gender, upbringing, and other factors. A voice can also have an *accent.*

You can probably tell when a person speaking to you is angry or happy, regardless of whether that person is from Texas, Indiana, Idaho, or Maine. However, some accents sound more authoritative than others; some sound funny if you have not been exposed to them before. Along with accent, the choice of word usage varies in different regions. This is *dialect*. For robotics engineers, producing a speech synthesizer with a credible "tone of voice" is a challenge.

Record and playback

The most primitive form of speech synthesizer is a set of tape recordings of individual words. You have heard these in automatic telephone answering machines and services. Most cities have a telephone number you can call to get local time; some of these are word recordings. They all have a characteristic choppy, interrupted sound.

There are several drawbacks to these systems. Perhaps the biggest problem is the fact that each word requires a separate recording, on a separate length of tape. These tapes must be mechanically accessed, and this takes time. It is impossible to have a large speech vocabulary using this method.

Reading text

Printed text can be read by a machine using *optical character recognition* (OCR), and converted into a standard digital code called ASCII (pronounced "ASK-ee"). The ASCII can be translated by a *digital-to-analog converter (DAC)* into voice sounds. In this way, a machine can read text out loud. Although they are rather expensive at the time of this writing, these machines are being used to help blind people read printed text.

Because there are only 128 allophones in the English language, a machine can be designed to read almost any text. However, machines lack a sense of which inflections are best for the different scenes that come up in a story. With technical or scientific text, this is rarely a problem, but in reading a story to a child, mental imagery is crucial. It is like an imaginary movie, and it is helped along by the emotions of the reader. No machine yet devised can paint pictures, or elicit moods, in a listener's mind as well as a human being. These things are apparent from context. The tone of a sentence might depend on what happened in the previous sentence, paragraph, or chapter. Technology is a long way from giving a machine the ability to understand, and appreciate, a good story, but nothing short of that level of AI will produce a vivid "story movie" in a listener's mind.

The process

There are several ways in which a machine can be programmed to produce speech. A simplified block diagram of one process is shown in the

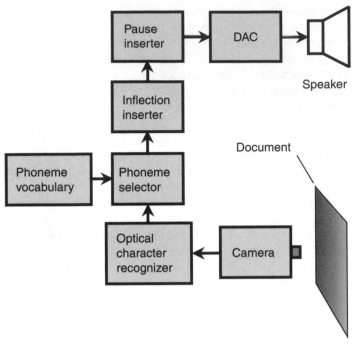

Speech synthesis

illustration. Whatever method is used for speech synthesis, certain steps are necessary. These are as follows:

- The machine must access the data and arrange it in the proper order.
- The allophones must be assigned in the correct sequence.
- The proper inflections must be put in.
- Pauses must be inserted in the proper places.

In addition to the foregoing, features such as the following can be included for additional versatility and realism:

- An intended mood can be conveyed (joy, sadness, urgency, etc.) at various moments.
- Overall knowledge of the content can be programmed in. For example, the machine can know the significance of a story, and the importance of each part within the story.
- The machine can have an interrupt feature to allow conversation with a human being. If the human says something, the machine will stop and begin listening with a *speech-recognition* system.

This last feature could prove interesting if two computers, both equipped with AI, speech synthesis, and speech recognition, got into an argument. One machine might be programmed as a Republican and the other as a Democrat; the engineer could bring up the subject of taxes and let the two machines argue.

For further information

The best source of up-to-date information on speech synthesis is a good college library. A search on the phrases "speech synthesis" and "voice synthesis" can be conducted on the Web using Google (www.google.com) or a similar search engine. Related entries include: BANDWIDTH, CONTEXT, DATA CONVERSION, DIGITAL SIGNAL PROCESSING, MESSAGE PASSING, OPTICAL CHARACTER RECOGNITION, PROSODIC FEATURES, SOUND TRANSDUCER, SPEECH RECOGNITION, and SYNTAX.

SPHERICAL COORDINATE GEOMETRY

Spherical coordinate geometry is a scheme for guiding a robot arm in three dimensions. A spherical coordinate system is something like the polar system, but with two angles instead of one. In addition to the two angles, there is a radius coordinate.

One angle, call it *x*, is measured counterclockwise from the reference axis. The value of *x* can range from 0° to 360°. You might think of *x* as

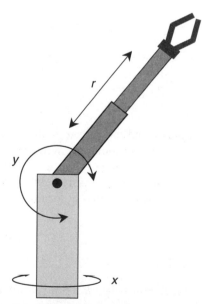

Spherical coordinate geometry

similar to the *azimuth* bearing used by astronomers and navigators, except that it is measured counterclockwise rather than clockwise. As a ray rotates around a full circle through all possible values of *x*, it defines a reference plane.

The second angle, call it *y*, is measured either upward or downward from the reference plane. The value of *y* will ideally range from −90° (straight down) to +90° (straight up). Structural limitations of the robot arm might limit the lower end of this range to something like −70°. You might think of *y* as the *elevation* above or below the horizon.

The radius, denoted *r*, is a non-negative real number (zero or greater). It can be specified in units such as centimeters, millimeters, or inches.

The illustration shows a robot arm equipped for spherical coordinate geometry. The movements *x*, *y*, and *r* are called *base rotation, elevation,* and *reach,* respectively. Compare CARTESIAN COORDINATE GEOMETRY, CYLINDRICAL COORDINATE GEOMETRY, POLAR COORDINATE GEOMETRY, and REVOLUTE GEOMETRY.

STADIMETRY

Stadimetry is a method that a robot can use to measure the distance to an object when the object's height, width, or diameter is known. The vision system and controller ascertain the angular diameter of the object. The linear dimension of the object must be known. The distance can then be calculated using trigonometry.

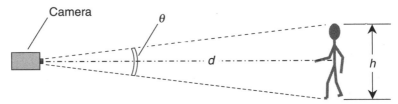

Stadimetry

The illustration shows an example of stadimetry as it might be used to measure the distance *d*, in meters, from a robot camera to a person. Suppose the person's height *h*, in meters, is known. The vision system determines the angle θ that the person subtends in the field of view. From this information, the distance *d* is calculated according to the following formula:

$$d = \frac{h}{2\tan(\theta/2)}$$

If the distance *d* is large compared with the height *h*, a simpler formula can be used:

$$d = \frac{h}{\tan \theta}$$

In order for stadimetry to be accurate, the linear dimension axis (in this case the axis that depicts the person's height, h) must be perpendicular to a line between the vision system and the center of the object. Also, it is important that d and h be expressed in the same units.

See also DISTANCE MEASUREMENT.

STATIC STABILITY

Static stability is the ability of a robot to maintain its balance while standing still. A robot with two or three legs, or that rolls on two wheels, usually has poor static stability. It might be all right as long as it is moving, but when it comes to rest, it can easily fall over. A bicycle is an example of a machine with good dynamic stability (it is all right while rolling), but poor static stability (it will not stand up by itself when at rest).

For a two-legged or three-legged robot to have excellent static stability, it needs a sense of balance. You can stand still and not fall over, because you have this sense. If your sense of balance is upset, you will topple. It is difficult to build a good sense of balance into a two-wheeled or two-legged robot. However, it has been done, and although the technology is expensive, it holds promise for the future.

See also BIPED ROBOT, DYNAMIC STABILITY, and INSECT ROBOT.

STEPPER MOTOR

A *stepper motor* is an electric motor that turns in small increments rather than continuously. Stepper motors are used extensively in robots.

Stepper versus conventional

When electric current is applied to a conventional motor, the shaft turns continuously at high speed. With a stepper motor, the shaft turns a little and then stops. The *step angle,* or extent of each turn, varies depending on the particular motor. It can range from less than 1° of arc to a quarter of a circle (90°).

Conventional motors run constantly as long as electric current is applied to the coils. A stepper motor turns through its step angle and then stops, even if the current is maintained. In fact, when a stepper motor is stopped with a current going through its coils, the shaft resists turning. A stepper motor has brakes built in. This is a great advantage in robotics; it keeps a robot arm from moving out of place if it is bumped accidentally.

Conventional motors run at hundreds or even thousands of revolutions per minute (rpm). A typical speed is 3600 rpm, or 60 revolutions per second (rps). A stepper motor, however, usually runs less than 180 rpm, or 3 rps. Often the speed is much slower than that. There is no lower limit; a robot arm might be programmed to move just 1° per day, if a speed that slow is necessary.

In a conventional motor, the *torque,* or turning force, increases as the motor runs faster. With a stepper motor, however, the torque decreases as the motor runs faster. Because of this, a stepper motor has the most turning power when it is running at slow speed. In general, stepper motors are less powerful than conventional motors.

Two-phase and four-phase

The most common stepper motors are of two types: two-phase and four-phase. A *two-phase stepper motor* has two coils, called *phases,* controlled by four wires. A *four-phase stepper motor* has four phases and eight wires. The motors are stepped by applying current sequentially to the phases. The illustration shows schematic diagrams of two-phase and four-phase stepper motors.

When a pulsed current is supplied to a stepper motor, with the current rotating through the phases, the motor rotates in steps, one step for each pulse. In this way, a precise speed can be maintained. Because of the braking effect, this speed is constant for a wide range of mechanical turning resistances. Most stepper motors can work with pulse rates up to about 200 per second.

Control

Stepper motors can be controlled using microcomputers. Several stepper motors, all under the control of a single microcomputer, are typical in robot arms of all geometries. Stepper motors are especially well suited for *point-to-point motion.* Complicated, intricate tasks can be done by robot arms using stepper motors controlled by software. The task can be changed by changing the software. This can be as simple as launching a new program with a spoken or keyed-in command. Compare SELSYN, SERVOMECHANISM, and SYNCHRO.

See also MOTOR, POINT-TO-POINT MOTION, and ROBOT ARM.

STEREOSCOPIC VISION

See BINOCULAR MACHINE VISION.

SUBMARINE ROBOT

Human SCUBA divers cannot normally descend to levels deeper than about 300 m (1000 ft). Rarely do they descend below 100 m (330 ft). Even at this depth, a tedious period of decompression is necessary to prevent

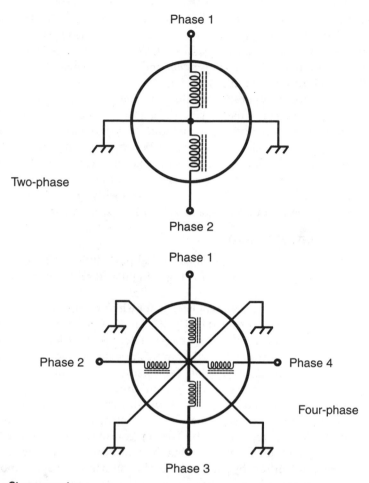

Phase 1

Phase 2

Two-phase

Phase 1

Phase 2   Phase 4

Four-phase

Phase 3

Stepper motor

illness or death from "the bends." Not surprisingly, there is great interest in developing robots that can dive down more than 300 m while doing all, or most, of the things that human divers can do.

The ideal submarine robot employs *telepresence*. This is an advanced form of remote control in which the operator has the impression of "being the robot."

Imagine a treasure-hunting expedition, in which you salvage diamonds, emeralds, and gold coins from a sunken galleon 1000 m beneath the surface of the sea, while sitting warm and dry in a remote-control chair. Imagine testing shark repellants without fear. Imagine disarming a sunken warhead at the bottom of a deep bay, or repairing a deep-sea observation station.

The remains of the *Titanic*, the "unsinkable" ocean liner that sank after colliding with an iceberg, were found and photographed by an undersea robot called a *remotely operated vehicle* (ROV). This machine did not feature telepresence, but it did use *teleoperation*, and it provided many high-quality pictures of the wrecked ship.

A specialized form of ROV is called an *autonomous underwater vehicle* (AUV). This machine has a cable through which control signals and response data pass. In underwater applications, radio control is not feasible because the water blocks the electromagnetic fields. The cable can use either electrical signals or fiber-optic signals. An alternative, wireless method of remotely controlling undersea robots is ultrasound. One type of AUV is called the *flying eyeball*. It is basically a camera with outboard motors.

See also FLYING EYEBALL, REMOTE CONTROL, TELEOPERATION, and TELEPRESENCE.

SURGICAL ASSISTANCE ROBOT

Robots have found a role in some surgical procedures. Robotic devices are steadier, and are capable of being manipulated more precisely, than any human hand.

Drilling in the skull is one application for which robots have been used. This technique was pioneered by Dr. *Yik San Kwo*, an electrical engineer at Memorial Medical Center in Southern California. The drilling apparatus is positioned by software derived from a computerized X-ray scan, called a *computerized axial tomography (CAT) scan*, of the brain. All of the robotic operation is overseen by a human surgeon.

Numerous applications for robots in surgery have been suggested. One of the more promising ideas involves using a teleoperated robot controlled by a surgeon's hands. The surgeon observes the procedure while going through the motions, but the actual contact with the patient is carried out entirely by the machine. Human hands always tremble a little. As a surgeon gains experience, he or she also grows older, and the trembling increases. The teleoperated robot would eliminate this problem, allowing surgeons with much experience (but limited dexterity) to perform critical operations.

See also TELEOPERATION.

SWARM ROBOT

See INSECT ROBOT.

SYNCHRO

A *synchro* is a special type of motor, used for remote control of mechanical devices. It consists of a generator and a receiver motor. As the shaft of the generator is turned, the shaft of the receiver motor follows along exactly.

In robots, synchros find many different uses. They are especially well suited to fine motion, and also to teleoperation. A simple synchro, used to indicate direction, is called a *selsyn*.

Some synchro devices are programmable. The operator inputs a number into the synchro generator, and the receiver changes position accordingly. Computers allow sequences of movements to be programmed. This allows complex, remote-control robot operation. Compare SELSYN, SERVOMECH-ANISM, and STEPPER MOTOR.

See also TELEOPERATION.

SYNTAX

Syntax refers to the way a sentence, either written or spoken, is put together. It is important in *speech recognition* and *speech synthesis*. It is also important in computer programming. Each high-level language has its own unique syntax.

You studied sentence structure in middle-school English grammar classes. Most students find it boring, but it can be fascinating if you have a good teacher. Diagramming sentences is like working with mathematical logic. Computers are good at this. Some engineers spend their careers figuring out new and better ways to interface human language with computers.

There are several basic sentence forms; all sentences can be classified into one of these forms. The sentence "John lifts the tray," for example, might be called SVO for *subject/verb/object*. "John" is the subject, "lifts" is the verb, and "tray" is the object.

Different languages have different syntax rules. In the Russian language, "I like you" is said as "I you like." That is, an SVO sentence is really SOV. The meaning is clear, as long as the syntax rules are known. However, if the syntax rules are not known, the meaning can be lost.

When designing a robot that can talk with people, engineers must program syntax rules into the controller. Otherwise the robot might make nonsensical statements, or misinterpret what people say.

See also CONTEXT, PROSODIC FEATURES, SPEECH RECOGNITION, and SPEECH SYNTHESIS.

T

TACTILE SENSING

The term *tactile sensing* refers to various electromechanical methods of simulating the sense of human touch. These include the abilities to sense and measure pressure, linear force, torque, temperature, and surface texture. Some roboticists consider tactile sensors second in importance only to vision systems.

The following entries contain information about tactile sensing and related subjects: BACK PRESSURE SENSOR, DISPLACEMENT TRANSDUCER, ELASTOMER, EYE-IN-HAND SYSTEM, FEEDBACK, FINE MOTION PLANNING, GRASPING PLANNING, JOINT-FORCE SENSING, POSITION SENSING, PRESSURE SENSING, PROPRIOCEPTOR, PROSTHESIS, TEMPERATURE SENSING, TEXTURE SENSING, and WRIST-FORCE SENSING.

TANGENTIAL FIELD

See POTENTIAL FIELD.

TASK ENVIRONMENT

The term *task environment* refers to the characteristics of the space where a robot, or a group of robots, operates. The task environment is also called the *world space*. The nature of the task environment depends on many factors that often interact. Some things that affect the task environment are

- The nature of the work the robot(s) must do
- The design of the robot(s)
- The speed at which the robot(s) work
- How many robots are in the area
- Whether or not humans work with the robot(s)
- Whether or not dangerous materials are present
- Whether or not any of the work is hazardous

An *autonomous robot* can benefit from a *computer map* of its task environment. This will minimize unnecessary movements, and will reduce the chance for mishaps, such as a robot falling down the stairs or crashing through a window.

When a robot is anchored down in one place, as are many industrial robots, the task environment is called the *work envelope.*

See also AUTONOMOUS ROBOT, COMPUTER MAP, and WORK ENVELOPE.

TASK-LEVEL PROGRAMMING

As machines become smarter, the programming becomes more sophisticated. No machine has yet been built that has intelligence anywhere near that of a human being. Some researchers think that true *artificial intelligence* (AI), at a level near that of the human brain, will never be achieved.

The programming of robots can be divided into levels, starting with the least sophisticated and progressing to the theoretical level of true AI. The drawing shows a four-level scheme. Level 3, just below AI, is called *task-level programming.* As the name implies, programs at this level encompass whole tasks, such as cooking meals, mowing a lawn, or cleaning a house.

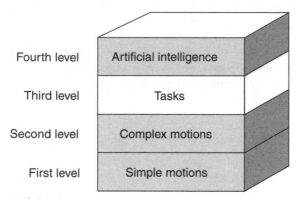

Fourth level	Artificial intelligence
Third level	Tasks
Second level	Complex motions
First level	Simple motions

Task-level programming

Task-level programming lies just above the hierarchy from complex-motion planning, but below the level of sophistication generally considered to be AI. Compare ARTIFICIAL INTELLIGENCE, COMPLEX-MOTION PROGRAMMING, and SIMPLE-MOTION PROGRAMMING.

TEACH BOX

When a robot arm must perform repetitive, precise, complex motions, the movements can be entered into the robot controller's memory. Then, when the memory is accessed, the robot arm goes through all the appropriate movements. A *teach box* is a device that detects and memorizes motions or processes for later recall.

In the four-level programming hierarchy shown in the drawing for TASK-LEVEL PROGRAMMING, the first and second levels are commonly programmed in teach boxes. In some cases, a primitive form of the third level can be programmed.

An example of a level-1 teach box is an automatic garage-door opener/closer. When the receiver gets the signal from the remote unit, it opens or closes the door. Another example of a level-1 teach box is the remote control that you use to change the channel and adjust the volume on a television set.

An example of a level-2 teach box is the microcomputer that controls a telephone answering machine. When a call comes in, the sequence of operations is recalled from the microcomputer memory. The machine answers the phone, makes an announcement, takes the message, and resets for the next incoming call.

Reprogrammable teach boxes are used extensively in industrial robots. The arm movements can be entered by pressing buttons. In some cases, it is possible to guide the robot arm manually (that is, "teach" it), and have the sequence of movements memorized for the performance of a specific task. The arm's path, variations in speed, rotations, and gripping/grasping movements are all programmed as the arm is "taught." Then, when the memory is recalled, the robot arm behaves just as it has been "taught."

See also FINE MOTION PLANNING, GROSS MOTION PLANNING, ROBOT ARM, and TASK-LEVEL PROGRAMMING.

TECHNOCENTRISM

During the twentieth century, people became increasingly comfortable with computers, machines, and electronic devices. This trend is expected to continue. Gadgets can be fascinating. *Technocentrism* refers to a keen interest in technology on the part of individuals, groups, and societies. In the extreme, of course, it can become an obsession.

Enthusiasm for technology can lead to exciting and rewarding careers, but if it goes too far, it can throw a person's life out of balance. Some technocentrics have difficulty relating to other human beings. Critics of technological advancement claim that the same thing is taking place in society as a whole. Technocentrism is a phenomenon that some sociologists believe has become a social disease.

Most people are familiar with the downside of technocentrism. People build and buy machines to make life simpler and more relaxed; but for some strange reason, their lives get more complicated and tense. People find themselves attending to machines that are more and more complex. The machines break down, and people must take them in for repair. Machines get more versatile, but people must learn to use the new features.

Rather than allowing us more free time, our technological miracles seem to devour our time and attention.

To avoid the less desirable effects of technocentrism, humanity must adopt a balanced outlook. Humans must be, and must always remain, the masters of machines.

See also UNCANNY VALLEY.

TELECHIR

See TELEOPERATION and TELEPRESENCE.

TELEMETRY

Telemetry is the transmission of quantitative information from one point to another, usually by wireless means, and particularly by radio. Telemetry is used extensively to monitor conditions in the vicinity of remote devices such as robots, weather balloons, aircraft, and satellites. Telemetry is used in space flights, both manned and unmanned, to keep track of all aspects of the equipment and the physical condition of astronauts.

A telemetry transmitter consists of a measuring instrument or set of instruments, an encoder that translates the instrument readings into electrical impulses, and a modulated radio transmitter with an antenna. A telemetry receiver consists of a radio receiver with an antenna, a demodulator, and a recorder. A computer is often used to process the data received. Data conversion might be necessary at either the transmitter end (the remotely controlled device or system), the receiver end (usually the station attended by a human operator), or both.

See also DATA CONVERSION, REMOTE CONTROL, TELEOPERATION, and TELEPRESENCE.

TELEOPERATION

Teleoperation is the technical term for the remote control of *autonomous robots*. A remotely controlled robot is called a *telechir*.

In a teleoperated robotic system, the human operator can control the speed, direction, and other movements of a robot from some distance away. Signals are sent to the robot to control it; other signals come back, telling the operator that the robot has followed instructions. The return signals are called *telemetry*.

Some teleoperated robots have a limited range of functions. A good example is a space probe, such as *Voyager*, hurtling past some remote planet. Earthbound scientists sent commands to *Voyager* based on the telemetry received from it, aiming its cameras and fixing minor problems. *Voyager* was, in this sense, a teleoperated robot.

Teleoperation is used in robots that can look after their own affairs most of the time, but occasionally need the intervention of a human operator.

See also AUTONOMOUS ROBOT, CONTROL TRADING, REMOTE CONTROL, SHARED CONTROL, and TELEPRESENCE.

TELEPRESENCE

Telepresence is a refined, advanced form of *teleoperation*. The robot operator gets a sense of being at the robot's physical location, even if the remotely controlled robot (or *telechir*) and the operator are many kilometers apart. Using *master–slave manipulators,* the robot can duplicate the movements of the operator. Control of these manipulators is accomplished by means of signals sent and received over wires, cables, optical fibers, or radio.

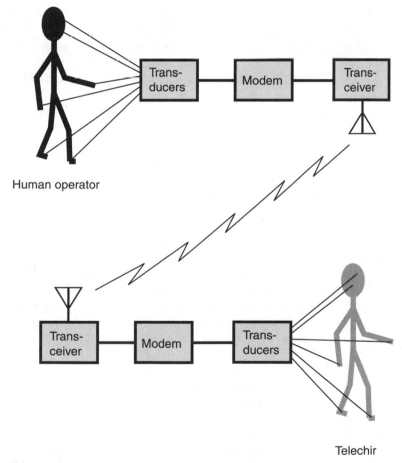

Human operator

Telechir

Telepresence

The drawing is a simple block diagram of a telepresence system. Some applications are

- Working in extreme heat or cold
- Working under high pressure, such as on the sea floor
- Working in a vacuum, such as in space
- Working where there is dangerous radiation
- Disarming bombs
- Handling toxic substances
- Law-enforcement operations
- Military operations

The experience

In a telepresence system, the robot is autonomous, and in some cases takes the physical form of a human body. The more humanoid the robot, the more realistic is the telepresence. The control station consists of a suit that the operator wears, or a chair in which the operator sits with various manipulators and displays. Sensors and transducers can impart feelings of pressure, vision, and sound.

In the most advanced telepresence systems, the operator wears a helmet with a viewing screen that shows whatever the robot camera sees. When the operator's head turns, the robot head, with its vision system, follows. Thus, the operator sees a scene that changes with turns of the head, replicating the effect of being on-site. Binocular robot vision can provide a sense of depth. Binaural robot hearing allows the perception of sounds.

The telechir can be propelled by a track drive, a wheel drive, or robot legs. If the propulsion uses legs, the operator can propel the robot by walking around a room. Otherwise the operator can sit in a chair and "drive" the robot like a cart.

A typical android telechir has two arms, each with grippers resembling human hands. When the operator wants to pick something up, he or she goes through the motions. Back pressure sensors and position sensors impart a sensation of heft. The operator might throw a switch, and something that weighs 10 kg will feel as if it only weighs 1 kg.

Limitations

The technology for advanced, realistic telepresence, comparable to virtual-reality experience, exists, but there are some difficult problems and challenges.

The most serious limitation is the fact that telemetry cannot, and never will, travel faster than the speed of light in free space. This seems fast at first thought (299,792 km/s, or 186,282 mi/s), but it is slow on an interplanetary scale. The Moon is more than a light-second away from Earth; the Sun is 8 light-minutes away. The nearest stars are at distances of several light-years. The delay between the sending of a command and the arrival of the return signal must be less than 0.1 s if telepresence is to be realistic. This means that the telechir cannot be more than about 15,000 km, or 9300 mi, away from the control operator.

Another problem is the resolution of the robotic vision system. A human being with good eyesight can see things with several times the detail of the best fast-scan television sets. To send that much detail, at realistic speed, requires a huge signal bandwidth. There are engineering problems (and cost problems) that go along with this.

Still another limitation is best put as a question: How will a robot be able to "feel" something and transmit these impulses to the human brain? For example, an apple feels smooth, a peach feels fuzzy, and an orange feels shiny yet bumpy. How can this sense of texture be realistically transmitted to the human brain? Will people allow electrodes to be implanted in their brains so they can perceive the universe as if they are robots?

For further information

For details about the latest progress in this field, consult a good college or university library. The Internet can be a useful source of information, but one must check the dates on which Web sites were last revised. Information on some specific topics can be found in this book under the following headings: AUTONOMOUS ROBOT, BACK PRESSURE SENSOR, BINOCULAR MACHINE VISION, BINAURAL MACHINE HEARING, CAPACITIVE PROXIMITY SENSING, EXOSKELETON, FLYING EYEBALL, MILITARY ROBOT, SURGICAL ASSISTANCE ROBOT, POLICE ROBOT, POSITION SENSING, PRESSURE SENSING, PROPRIOCEPTOR, REMOTE CONTROL, ROBOTIC SPACE TRAVEL, SUBMARINE ROBOT, TACTILE SENSING, TELEOPERATION, TEMPERATURE SENSING, TEXTURE SENSING, and VISION SYSTEM.

TEMPERATURE SENSING

In a robotic system, *temperature sensing* is one of the easiest things to do. Digital thermometers are commonplace nowadays, and cost very little. The output from a digital thermometer can be fed directly to a micro-computer or robot controller, allowing a robot to ascertain the temperature at any given location.

Temperature data can cause a robotic system to behave in various ways. An excellent practical example is a fleet of *fire-protection robots*. Temperature sensors can be located in many places throughout a house, manufacturing plant, nuclear power plant, or other facility. At each sensor location, a critical temperature can be determined in advance. If the temperature at some sensor location rises above the critical level, a signal is sent to a central computer. The computer can dispatch one or more robots to the scene. These robots can determine the source and nature of the problem, and take action.

See also FIRE-PROTECTION ROBOT, THERMISTOR, and THERMOCOUPLE.

TETHERED ROBOT

A *tethered robot* is a semimobile robot that receives its commands from, and transmits its data to, the controller through a cable. The cable can be of the "copper" variety, which transmits electrical signals, or the fiber-optic variety, which transmits infrared (IR) or visible-light signals. The cable serves the dual purpose of transmitting data and keeping the machine from wandering outside a prescribed work environment.

Tethered robots are used in scenarios where wireless modes are impractical or difficult to use. A good example is submarine observation, particularly the exploration of underwater caverns or shipwrecks.

See also FLYING EYEBALL.

TEXTURE SENSING

Texture sensing is the ability of a robot *end effector* to ascertain the smoothness or roughness of a surface. Primitive texture sensing can be done with a laser and several light-sensitive sensors.

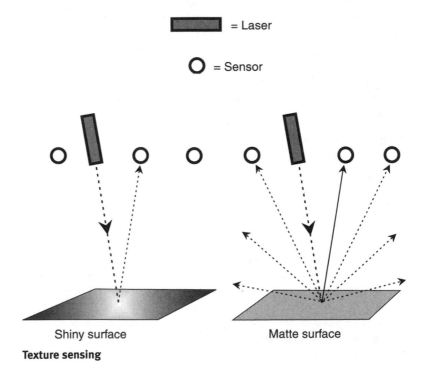

Texture sensing

The illustration shows how a laser (dark rectangle) can be used to tell the difference between a shiny surface (at left) and a rough or matte surface (at right). A shiny surface, such as the polished hood of a car, tends to reflect light according to the rule of reflection, which states that the angle of reflection equals the angle of incidence. A matte surface, such as the surface of a sheet of drawing paper, scatters the light. The shiny surface reflects the beam back almost entirely to one of the sensors (circles), positioned in the path of the beam whose reflection angle equals its incidence angle. The matte surface reflects the beam back more or less equally to all of the sensors.

The visible-light texture sensing scheme cannot give an indication of relative roughness. It can only let a robot know that a surface is either shiny, or not shiny. A piece of drawing paper reflects the light in much the same way as a sandy beach or a new-fallen layer of snow. The measurement of relative roughness, or of the extent to which a grain is coarse or fine, requires more sophisticated techniques.

See also TACTILE SENSING.

THERMISTOR

A *thermistor* is an electronic sensor designed specifically so that its resistance changes with temperature. The term thermistor is a contraction of "thermally sensitive resistor."

Thermistors are made from semiconductor materials. The most common substances used are oxides of metals. In some thermistors, the resistance increases as the temperature rises; in others, the resistance decreases as the temperature rises. In either type of thermistor, the resistance is a precise function of the temperature.

Thermistors are used for *temperature sensing* and measurement. The resistance-versus-temperature characteristic makes the thermistor ideal for use in thermostats and thermal protection circuits. Thermistors are operated at low current levels, so that the resistance is affected only by the ambient temperature, and not by heating caused by the applied current itself. Compare THERMOCOUPLE.

See also TEMPERATURE SENSING.

THERMOCOUPLE

A *thermocouple* is an electronic sensor designed to facilitate the measurement of temperature differences. The device consists of two wires or strips of specially chosen dissimilar metals, such as antimony and bismuth, placed in contact with each other.

When the two metals are at the same temperature, the voltage between them is zero. However, when the metals are at different temperatures, a direct-current (DC) voltage appears between them. The magnitude of this voltage is directly proportional to the temperature difference within a limited range. The function of voltage in terms of temperature difference can be programmed into a robot controller, allowing the machine to determine the temperature difference by measuring the voltage. Compare THERMISTOR.

See also TEMPERATURE SENSING.

THREE LAWS OF ROBOTICS

See ASIMOV'S THREE LAWS.

TIME-OF-FLIGHT DISTANCE MEASUREMENT

Time-of-flight distance measurement, also called *time-of-flight ranging,* is a common method by which a robot can determine the straight-line distance between itself and an object. A wave or signal pulse, which travels at a known, constant speed, is transmitted from the robot. This signal reflects from the object, and a small amount of energy returns to the robot. The distance to the object is calculated on the basis of the time delay between the transmission of the original signal pulse and the reception of the return signal, or *echo.*

Suppose the speed of the signal disturbance, in meters per second, is denoted c, and the time delay, in seconds, is denoted t. Then the distance d to the object under consideration, assuming the signal travels through the same medium throughout the span between the robot and the object, is given by this formula:

$$d = \frac{ct}{2}$$

Examples of systems that use time-of-flight ranging are *ladar, radar,* and *sonar.* These use laser beams, microwave radio signals, and acoustic waves, respectively. The speed of laser beams or radio waves in Earth's atmosphere is approximately 300 million (3.00×10^8) m/s; the speed of acoustic waves in air at sea level is approximately 335 m/s.

See also LADAR, RADAR, and SONAR.

TIME SHIFTING

In communications, *time shifting* refers to any system in which there is a significant delay between the transmission of a signal at the source and its receipt or utilization at the destination. The term applies especially to computer networks and remotely controlled robotic systems. Time shifting does not allow a computer or robot and the operator to converse, but commands and telemetry can be conveyed.

Time shifting is best suited for the transmission of data at high speed, and in large blocks. This is the case, for example, when monitoring conditions at a far-off space probe. In computer systems and networks, time shifting can save expensive on-line time to write long programs or compose long messages at an active terminal. Compare REAL TIME.

TOPOLOGICAL PATH PLANNING

Topological path planning, also called *topological navigation,* is a scheme in which a robot can be programmed to negotiate its work environment. The method makes use of specific points called *landmarks* and *gateways,* along with periodic instructions for action.

Topological path planning is used by people in everyday life. Suppose you are in an unfamiliar town, and you need to find the library. You ask someone at a small corner grocery how to get to the library. The person says, while pointing in a certain direction, "Go down this street here until you get to the sugar mill. Turn left at the sugar mill. You will pass three traffic lights and then the road will bear left. Keep going around the curve to the left. Just as the curve ends, turn right and follow the bumpy street until you get to a red brick building with white window trim. The building will be on the right side of the road. That is the library. If you get to a large shopping mall on the left, you have gone too far; turn around and go back. The library will then, of course, be on the left-hand side of the road."

Topological path planning is a qualitative scheme. Note that in the above set of directions, specific distances are not indicated. If you follow the directions, however, you will reach the library, and a computer-controlled robot would find it as well. The instructions, although they do not contain information about specific distances and compass directions, nevertheless provide sufficient information to allow you (or the robot) to find the intended destination.

Topological path planning does not always work. In complex environments, or in environments that change geometry often, more sophisticated navigational schemes are required. Compare GRAPHICAL PATH PLANNING and METRIC PATH PLANNING.

See also COMPUTER MAP, GATEWAY, LANDMARK, and RELATIONAL GRAPH.

TRACK-DRIVE LOCOMOTION

When neither wheels nor legs effectively propel a robot over a surface, *track-drive locomotion* sometimes works. Track drive is used in military tanks, and in some construction vehicles.

A track drive has several wheels and a pair of belts or tracks, as shown in the illustration. (This drawing shows only one side of the track drive. An identical wheel-and-belt set exists on the other side, out of sight in this perspective.) The track can be rubber if the vehicle is small; metal is better for large, heavy machines. The track can have ridges or a tread on the outside; this helps it grip dirt or sand.

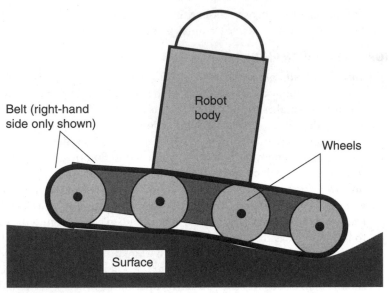

Track-drive locomotion

Assets

Track-drive locomotion works well in terrain strewn with small rocks. It is also ideal when the surface is soft or sandy. Track drive is often the best compromise for a machine that must navigate over a variety of different surfaces.

A special advantage of track drive is that the wheels can be suspended individually. This helps maintain traction over stones and other obstructions. It also makes it less likely that a moderate-sized rock will tip the robot over.

Steering is harder with track-drive than with wheel-drive locomotion. If the robot must turn right, the left-hand track must run faster than the right-hand track. If the robot is to turn left, the right-hand track must run faster than the left-hand track. Steering radius depends on the difference in speed between the two tracks.

Track drives can allow robots to climb or descend stairways, but for this to work, the track must be longer than the spacing between the stairs. Also, the whole track-drive system must be able to tilt up to 45°, while the robot remains upright. Otherwise the robot will fall backwards when going up the stairs, or forwards when going down. A better system for dealing with stairways is *tri-star wheel locomotion*.

Limitations

One potential problem with track drives is that the track can work its way off the wheels. The chances of this are reduced by proper wheel and track design. The inside surface of the track can have grooves, into which the wheels fit; or the inside of the track can have lip edges. The track must be wrapped snugly around the wheels. Some provision must be made to compensate for expansion and contraction of the belt with extreme changes in temperature.

Another problem with track drive is that the wheels might slip around inside the track, without the track following along. This is especially likely when the robot is climbing a steep slope. The machine will sit still or roll backwards despite the fact that its wheels are turning forwards. This can be prevented by using wheels with teeth that fit in notches on the inside of the track. The track then resembles a gear-driven conveyor belt.

On smooth surfaces, track drives are usually not needed. If a surface is extremely rugged, *robot legs* or *tri-star wheel locomotion* generally work better than wheels or track drives.

See also ADAPTIVE SUSPENSION VEHICLE, ROBOT LEG, TRI-STAR WHEEL LOCOMOTION, and WHEEL-DRIVE LOCOMOTION.

TRANSDUCER

A *transducer* is a device that converts one form of energy or disturbance into another. In electronics, transducers convert alternating or direct electric current into sound, light, heat, radio waves, or other forms. Transducers also convert sound, light, heat, radio waves, or other energy forms into alternating or direct electric current.

Common examples of electrical and electronic transducers include buzzers, speakers, microphones, piezoelectric crystals, light-emitting and infrared-emitting diodes, photocells, radio antennas, and many other devices.

In robotics, transducers are used extensively. For details on specific devices and processes, see BACK PRESSURE SENSOR, CAPACITIVE PROXIMITY SENSING, CHARGE-COUPLED DEVICE, CLINOMETER, DISTANCE MEASUREMENT, DISPLACEMENT TRANSDUCER, DYNAMIC TRANSDUCER, ELASTOMER, ELECTRIC EYE, ELECTROMECHANICAL TRANSDUCER, ELECTROSTATIC TRANSDUCER, ERROR-SENSING CIRCUIT, FLUXGATE MAGNETOMETER, JOINT FORCE SENSOR, ODOMETRY, OPTICAL CHARACTER RECOGNITION, OPTICAL ENCODER, PASSIVE TRANSPONDER, PHOTOELECTRIC PROXIMITY SENSING, POSITION SENSING, PRESSURE SENSING, PROPRIOCEPTOR, PROXIMITY SENSING, RANGE SENSING AND PLOTTING, SMOKE DETECTION, SONAR, SOUND TRANSDUCER, TACTILE SENSING, TEMPERATURE SENSING, TEXTURE SENSING, VISION SYSTEM, and WRIST-FORCE SENSOR.

TRIANGULATION

Robots can navigate in various ways. One good method is like the scheme that ship and aircraft captains have used for decades. It is called *triangulation.*

In triangulation, the robot has a direction indicator such as a compass. It also has a laser scanner that revolves in a horizontal plane. There must be at least two targets, at known, but different, places in the work environment, which reflect the laser beam back to the robot. The robot also has a sensor that detects the returning beams. Finally, it has a microcomputer that takes the data from the sensors and the direction indicator, and processes it to get its exact position in the work environment.

The direction sensor (compass) can be replaced by a third target. Then there are three incoming laser beams; the robot controller can determine its position according to the relative angles between these beams.

For optical triangulation to work, it is important that the laser beams not be blocked. Some environments contain numerous obstructions, such as stacked boxes, which interfere with the laser beams and make triangulation impractical. If a magnetic compass is used, it must not be fooled by stray magnetism; also, Earth's magnetic field must not be obstructed by metallic walls or ceilings.

The principle of triangulation, using a direction sensor and two reflective targets, is shown in the illustration. The laser beams (dashed lines) arrive from different directions, depending on where the robot is located with respect to the targets. The targets are *tricorner reflectors* that send all light rays back along the path from which they arrive.

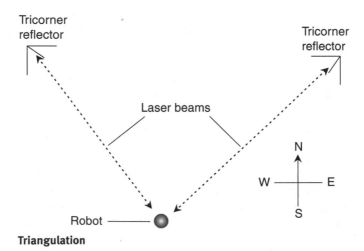

Triangulation

Triangulation need not use laser beams. Instead of the reflecting targets, beacons can be used. Instead of visible light, radio waves or sound waves can be used. Beacons eliminate the need for the 360° scanning transmitter in the robot.

See also BEACON, DIRECTION FINDING, DIRECTION RESOLUTION, LADAR, RADAR, and SONAR.

TRI-STAR WHEEL LOCOMOTION

A unique and versatile method of robot propulsion uses sets of wheels arranged in triangles. The geometry of the wheel sets has given rise to the term *tri-star wheel locomotion*. A robot can have three or more pairs of tri-star wheel sets. The illustration shows a robot with two sets. (This drawing shows only one side of the machine. An identical pair of tri-star wheels exists on the other side, out of sight in this perspective.)

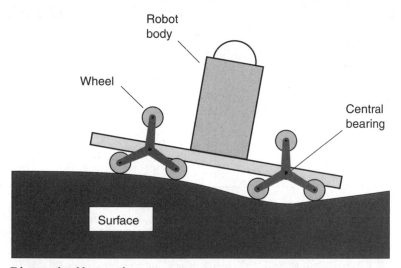

Tri-star wheel locomotion

Each tri-star set has three wheels. Normally, two of these are in contact with the surface. If the robot encounters an irregularity in the terrain, such as a big pothole or a field of rocks, the tri-star set rotates forward on a central bearing. Then, for a moment, only one of the three wheels is in contact with the surface. This might happen once or repeatedly, depending on the nature of the terrain. The rotation of the central bearing is independent of the rotation of the wheels.

Tri-star wheel locomotion works well for stair climbing. It can even allow a robot to propel itself through water, although slowly. The scheme was originally designed and patented by *Lockheed Aircraft*. Tri-star wheel locomotion is applicable to use by remotely controlled robots on the Moon or on distant planets.

See also ADAPTIVE SUSPENSION VEHICLE, ROBOT LEG, TRACK-DRIVE LOCOMOTION, and WHEEL-DRIVE LOCOMOTION.

TRUTH TABLE

A *truth table* is a way of breaking down a logical expression. Truth tables show outcomes for all possible situations. The table shows an example. It is arranged in columns, with each column representing some part of the whole expression. Truth values can be shown by T or F (true or false); often these are written as 1 and 0.

Truth table: breakdown of a logical statement

X	Y	Z	X + Y	− (X + Y)	XZ	− (X + Y) + XZ
0	0	0	0	1	0	1
0	0	1	0	1	0	1
0	1	0	1	0	0	0
0	1	1	1	0	0	0
1	0	0	1	0	0	0
1	0	1	1	0	1	1
1	1	0	1	0	0	0
1	1	1	1	0	1	1

The left-most columns of the truth table give the combinations of values for the inputs. This is done by counting upwards in the binary number system from 0 to the highest possible number. For example, if there are two variables, X and Y, there are four value combinations: 00, 01, 10, and 11. If there are three variables, X, Y, and Z, there are eight combinations: 000, 001, 010, 011, 100, 101, 110, and 111. If there are n variables, where n is a positive integer, then there exist 2^n possible truth combinations. When there are many variables, a truth table can become gigantic, especially when the logical expression is complex. Computers are ideal for working with such tables.

In most truth tables, X AND Y is written XY or X*Y. NOT X is written with a line or tilde over the quantity, or as a minus sign followed by the quantity. X OR Y is written X + Y.

The table shows a breakdown of a three-variable expression. All expressions in electronic logic, no matter how complicated, can be mapped in this way. Some people believe that the smartest machine, and even the human brain, works according to two-valued logic. If that is true, our brains are nothing more than massive sets of biological truth tables whose values constantly shift as our thoughts wander.

See also BOOLEAN ALGEBRA.

TURING TEST

The *Turing test* is one method that has been used in an attempt to find out if a machine can think. It was invented by logician Alan Turing.

The test is conducted by placing a male human (M), female human (F), and questioner (Q) in three separate rooms. None of the people can see the others. The rooms are soundproof, but each person has a video display terminal. In this way, the people can communicate. The object: Q must find out which person is male and which is female, on the basis of questioning them. But M and F are not required to tell the truth. Both M and F are told in advance that they may lie. The man is encouraged to lie often, and to any extent he wants. It is the man's job to mislead the questioner into a wrong conclusion. Obviously, this makes Q's job difficult, but the test is not complete until Q decides which room contains the man, and which contains the woman.

Suppose this test is done 1000 times, and Q is right 480 times and wrong 520 times. What will happen if the man is replaced by a computer, programmed to lie occasionally as does the man? Will Q be right more often, less often, or the same number of times as with the real man in the room?

If the man-computer has a low level of *artificial intelligence* (AI), then according to Turing's hypothesis, Q will be correct more often than when a human male is at the terminal, say 700 times out of 1000. If the man-computer has AI at a level comparable to that of a human male, Q should be right about the same number of times as when the man was at the terminal—say, correct 490 times and wrong 510 times. If the computer has AI at a level higher than that of the human male, then Q ought to be wrong most of the time—say, correct 400 times and mistaken 600 times.

The Turing test has not produced comprehensive results concerning computers with levels of AI higher than the intelligence of a human being, because no such computer has yet been developed. However, computers have been devised with high-level AI relative to specialized skills or tasks, such as board games. Powerful computers have proven equal to human masters at the game of chess, for example.

TWO-PINCHER GRIPPER

One of the simplest types of robot gripper uses two tongs or pinchers. Because of its construction, it is called a *two-pincher gripper*.

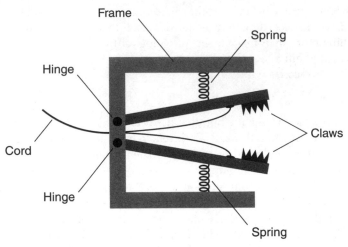

Two-pincher gripper

The drawing shows a simple version of a two-pincher gripper. The claws are attached to a frame, and are normally held apart by springs. The claws are pulled together by means of a pair of cords that merge into a single cord as shown. This allows the gripper to pick up small, light objects. To release the grip, the cord is let go.

See also ROBOT GRIPPER.

ULTRASONIC DIRECTION FINDER

See DIRECTION FINDING.

UNCANNY VALLEY THEORY

Some people are fond of the idea of building *androids,* or robots in the human image. But at least one roboticist, *Masahiro Mori,* has stated a belief that the "humanoid" approach to robot building is not necessarily always the best. If a robot gets too much like a person, Mori thinks, it will seem uncanny, and people will have trouble dealing with it.

Reactions to robots

According to Mori's notion, which he calls the *uncanny valley theory,* the more a robot resembles a human being, the more comfortable people are with the machine, to a point. When machines become too human-like, however, disbelief and unease set in. People get intimidated by, and in some cases afraid of, such robots.

Mori drew a hypothetical graph to illustrate his theory (see the illustration). The curve has a dip, or "valley," in a certain range where people get uneasy around robots. Mori calls this the *uncanny valley.* No one knows exactly how human-like a robot must become to enter this zone. It can be expected to vary depending on the type of robot, and also on the personality of the robot user or operator.

Intimidated by intelligence

A similar curve apparently applies to powerful computers. Some people have problems with personal computers. These people can usually work with pocket calculators, adding machines, cash registers, television remote controls, and the like; but when they sit down in front of a computer, they freeze up. This is called *cyberphobia* ("fear of computers").

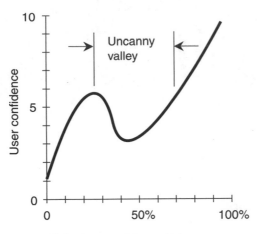

Uncanny valley theory

While some people are so intimidated by computers that they get a mental block right from the start, others are comfortable with them after a while, and only have problems when they try something new. Still other people never have any trouble at all.

The uncanny valley phenomenon is a psychological hangup that some people have with advanced technology of all kinds. A little skepticism is healthy, but overt fear serves no purpose and can keep a person from taking advantage of the good things technology has to offer. Some researchers think the uncanny valley problem can be avoided by introducing new technologies gradually. However, technology often seems to introduce itself at its own pace.

See also **ANDROID**.

UNIFORM POTENTIAL FIELD

See **POTENTIAL FIELD**.

UPLINK/DOWNLINK

In a system of mobile robots controlled by wireless means, the *uplink* is the frequency or band on which an individual robot receives its signals from a central controller. The uplink frequency or band is different from the *downlink* frequency or band, on which the robot transmits signals back to the controller. The terms uplink and downlink are used especially when a robot is a space satellite or probe with the controller located on Earth or on a space station.

In a fleet of *insect robots*, many mobile units receive and retransmit signals at the same time. For this to be possible, the uplink and downlink frequency bands must be substantially different. Also, the uplink band should not be harmonically related to the downlink band. The receiver must be well designed so it is relatively immune to *desensitization* and *intermodulation* effects. The receiving and transmitting antennas should be aligned in such a way that they are electromagnetically coupled to the least extent possible.

V

VACUUM CUP GRIPPER

A *vacuum cup gripper* is a specialized robotic *end effector* that employs suction to lift and move objects. The device consists of the gripper mechanism itself, a hose, an air pump, a power supply, and a connection to the robot controller.

In order to move an object from one location to another, the robot arm places the cup-shaped, flexible gripper against the surface of the object, which must be clean and nonporous so that air cannot leak around the edges of the cup. Then the robot controller actuates the motor, creating a partial vacuum inside the hose and cup assembly. Next, the robot arm moves the gripper from the *initial node* (starting position) to the intended *goal node* (final position). Then the robot controller briefly reverses the motor, so the pressure inside the hose and cup assembly returns to normal atmospheric pressure. Finally, the robot arm moves the gripper away from the object.

The principal asset of the vacuum cup gripper is the fact that it does not allow objects to slide out of position as they are moved. However, this type of gripper operates at limited speed, and is not capable of handling massive objects safely.

See also **ROBOT GRIPPER.**

VIA POINT

The term *via point* refers to any point through which a robot *end effector* passes as a manipulator moves it from the *initial node* (starting position) to the *goal node* (final position).

In the case of a robot arm that employs *continuous-path motion*, there are theoretically an infinite number of via points. In the case of *point-to-point motion*, the via points are those points at which the end effector can be halted; this set of points is finite.

See also **CONTINUOUS-PATH MOTION** and **POINT-TO-POINT MOTION.**

VIDEO SIGNAL

See COMPOSITE VIDEO SIGNAL.

VIDICON

Video cameras use a form of electron tube that converts visible light into varying electric currents. One common type of camera tube is called the *vidicon*. The illustration is a simplified, functional, cutaway view of a vidicon.

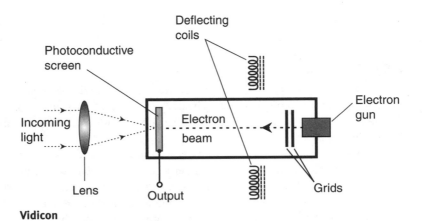

Vidicon

The camera in a common *videocassette recorder (VCR)* uses a vidicon. Closed-circuit television systems, such as those in stores and banks, also employ vidicons. The main advantage of the vidicon is its small physical bulk; it is easy to carry around. This makes it ideal for use in mobile robots.

In the vidicon, a lens focuses the incoming image onto a photoconductive screen. An electron beam, generated by an *electron gun,* scans across the screen in a pattern of horizontal, parallel lines called the *raster.* As the electron beam scans the photoconductive surface, the screen becomes charged. The rate of discharge in a certain region on the screen depends on the intensity of the visible light falling on that region. The scanning in the vidicon is exactly synchronized with the scanning in the display that renders the image on the vidicon screen.

A vidicon is sensitive, so it can see things in dim light. But the dimmer the light gets, the slower the vidicon responds to changes in the image. It gets "sluggish." This effect is noticeable when a VCR is used indoors at night. The image persistence is high under such conditions, and the resolution is comparatively low. Compare CHARGE-COUPLED DEVICE and IMAGE ORTHICON.

See also VISION SYSTEM.

VIRTUAL REALITY

Virtual reality (VR) is the ultimate simulator. The user can see and hear in an artificial realm called a *virtual universe* or *VR universe*. In the most sophisticated VR systems, other senses are replicated as well. Hardware and software developers in several countries, particularly the United States and Japan, are actively involved in VR technology.

Forms of VR

There are three degrees, or types, of VR. They are categorized according to the extent to which the operator participates in the experience. The first two forms are sometimes called *virtual virtual reality* (VVR).

Passive VR is, in effect, a movie with enhanced graphics and sound. You can watch, listen, and feel the show, but you have no control over what happens, nor on the general contents of the show. An example of passive VR is a ride in a virtual submarine, a small room with windows through which you can look at a rendition of the undersea world.

Exploratory VR is like a movie over which you have some control of the contents. You can choose scenes to see, hear, and feel, but you cannot participate fully in the experience. An example of exploratory VR is a ride in a tour bus on an alien planet, in which you get to choose the planet.

Interactive VR is what most people imagine when they think of true VR. You have nearly as much control over the virtual environment as you would have if you were really there. Your surroundings react directly to your actions. If you reach out and push a virtual object, it moves. If you speak to virtual people, they respond.

Programming

The program, or set of programs, containing all the particulars for each VR session is called the *simulation manager*. The complexity of the simulation manager depends on the level of VR.

One dimension: In passive VR, the simulation manager consists of a large number of frames, one representing each moment in time. The frames blend together into a space-time *experience path*. This can be imagined, in simplified form, as a set of points strung out along a straight line in one geometric dimension (Fig. 1). Each point represents data for one instant of time in the VR session. This is similar to the way frames exist in a movie or a videotape.

Two dimensions: In exploratory VR, there are several different sets of frames, from among which you can choose to construct the experience path. Imagine each set of frames as lying along its own individual line, as shown in Fig. 2. You choose the line through space-time along which

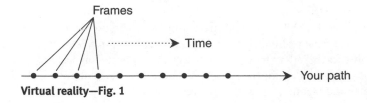

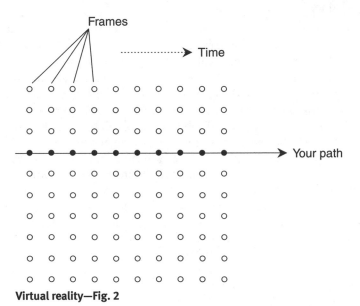

Virtual reality—Fig. 1

Virtual reality—Fig. 2

you want to travel. (Again, this is a simplistic rendition; there are far more points in an actual exploratory VR session than are shown here.) This is similar to having a selection of movies or videotapes from which to choose.

Three dimensions: In interactive VR, the sequence of frames depends on your input from moment to moment, adding another dimension to the programming. This can be rendered as a three-dimensional space (Fig. 3). The drawing shows only a few points along one path. There can be millions upon millions of points in the interactive experience space. The number of possible experience paths is vastly larger than the number of points themselves. It is impossible to make a good analogy with movies or videotapes in this case. The software, and the required computer hardware, for interactive VR is far more powerful than that in the passive or exploratory VR experiences.

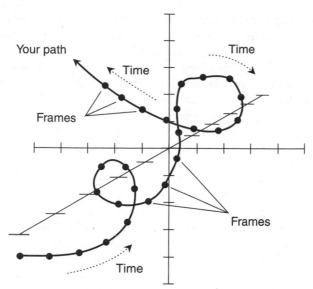

Your path

Time

Time

Frames

Frames

Time

Virtual reality—Fig. 3

Hardware

Several hardware items, in addition to the programming, are required for VR.

Computer: For VR to be possible, even in the simplest form, a computer is necessary. The amount of computer power required depends on the sophistication of the VR session. Passive VR requires the least computer power, while exploratory VR needs more, and interactive VR takes even more. A high-end personal computer can provide passive and exploratory VR with moderate image resolution and speed. Larger computers, such as those used in file servers or that employ *parallel processing* (more than one microprocessor operating on a given task), are necessary for high-resolution, high-speed, and vivid interactive VR. The best interactive VR equipment is too expensive for most personal-computer users.

Robot: If the VR is intended to portray and facilitate the operation of a remotely controlled robot or *telechir,* that robot must have certain characteristics. In low-level VR, the telechir can be a simple vehicle that rolls on wheels or a track drive. In the most sophisticated VR *telepresence* systems, the telechir must be an *android* (humanoid robot).

Video system: This can be a simple monitor, a big screen, a set of several monitors, or a *head-mounted display (HMD).* The HMD gives a spectacular show, with binocular vision and sharp colors. Some HMDs shut out the operator's view of the real world; others let the operator see the virtual universe superimposed on the real one. The HMD uses small liquid-crystal display (LCD) screens, whose images are magnified by lenses and/or reflected by mirrors to obtain the desired effects.

Sound system: Stereo, high-fidelity sound is the norm in all VR universes. Loudspeakers can be used for low-level, group VR experiences. In an individual system, a set of headphones is included in the HMD. The sound programming is synchronized with the visual programming. *Speech recognition* and *speech synthesis* can be used so that virtual people, virtual robots, or virtual space aliens can communicate their virtual thoughts and feelings to the user.

Input devices: Passive and exploratory VR systems do not need input devices, except for the media that contain the programming. Interactive systems can make use of a variety of mechanical equipment. The nature of the input device(s) depends on the VR universe. For example, driving a car requires a steering wheel, gas pedal, and brake (at least). Games need a joystick or mouse. Devices called *bats* and *birds* resemble computer mice, but are movable in three dimensions rather than only two. Levers, handles, treadmills, stationary bicycles, pulley weights, and other devices allow for real physical activities on the part of the operator. For complete hand control, special gloves can be used. These have air bladders built in, providing a sense of touch and physical resistance, so objects seem to have substance and heft. The computer can be equipped with speech recognition and speech synthesis so that the user can talk to virtual creatures. This requires at least one *sound transducer* at the operating location.

A complete system: Figure 4 is a block diagram showing the hardware for a typical interactive VR system, in which the user gets the impression of riding a bicycle down a street. This can be used for exercise as well as for entertainment. The system provides sights, sounds, and variable pedal resistance as the user negotiates hills and encounters wind. If a tele-operated android is put on a real bicycle, the VR system can be used to control that robot and bicycle remotely. This would involve the addition of two wireless transceivers (one for the computer and the other for the telechir), along with modems and antennas.

Applications

Virtual reality has been used as an entertainment and excitement medium. It also has practical applications.

Instruction: Virtual reality can be used in *computer-assisted instruction (CAI).* For example, a person can be trained to fly an aircraft, pilot a submarine, or operate complex and dangerous machinery, without any danger of being injured or killed during training. This form of CAI has been used by the military for some time. It has also been used for training

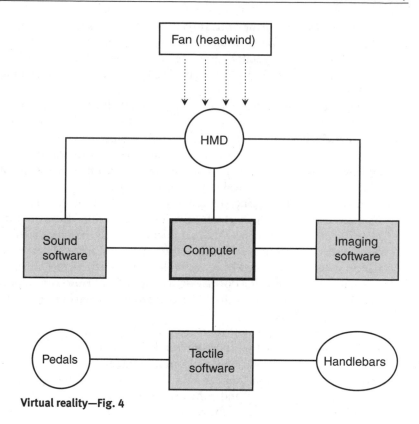

Virtual reality—Fig. 4

medical personnel, particularly surgeons, who can operate on "virtual patients" while they perfect their skills.

Group VR: Passive and exploratory VR can be provided to groups of people. Several theme parks in the United States and Japan have already installed equipment of this type. People sit in chairs while they watch, and listen to, the portrayal of an intergalactic journey, a submarine ride, or a trip through time. The main limitation is that everyone has the same virtual experience.

Individual VR: Interactive VR, intended for individual users, is also found in theme parks. These sessions are expensive and generally last only a few minutes. You might walk on an alien planet inhabited by robots, ride in a Moon buggy, or swim with porpoises. The environment reacts to your input from moment to moment. You might go through the same 10-min "show" 100 times and have 100 different VR experiences.

Hostile environments: In conjunction with robotics, VR facilitates remote control using *telepresence*. This allows a human operator to safely operate machines located in dangerous places. People using such a system get illusions similar to those in theme parks, except that a robot, at some distance, follows the operator's movements. Teleoperated robots have been used for rescue operations, for disarming bombs, and for maintaining nuclear reactors.

Warfare: Teleoperated robot tanks, aircraft, boats, and androids (humanoid robots) can be used in combat. One person can operate a "super android" with the strength of 100 fighting men and the endurance of a well-engineered machine. Such robots are immune to deadly radiation and chemicals. They have no mortal fear, which sometimes causes human soldiers to freeze up at critical moments in combat.

Exercise: Walking, jogging, riding a bike, skiing, playing golf, and playing handball are examples of virtual activities that can provide most of the benefits of the real experience. The user might not really be doing the thing, but calories are burned, and aerobic benefits are realized. There is no danger of getting maimed by a car while cycling on a virtual street, or breaking a leg while skiing down a virtual mountain. (However, outdoor people will doubtless prefer the real activity to the virtual one, no matter how realistic VR becomes.)

Escape: Another possible, but not yet widely tested, use for virtual reality is as an escape from boredom and frustration in the real world. You can put on an HMD and romp in a jungle with dinosaurs. If the monsters try to eat you, you can take the helmet off. You can walk on some unknown planet, or under the sea. You can fly high above clouds or tunnel through the center of the earth.

VR and space exploration

See ROBOTIC SPACE TRAVEL.

Limitations

The field of VR is complex, challenging, and difficult from an engineering standpoint. Dreaming up uses and scenarios for VR is one thing; putting them into action at reasonable cost is quite another.

Expense: A top-notch, interactive VR system can cost as much as $250,000. While a high-end personal computer and peripherals, costing about $5000 total, can be used for interactive VR, the image resolution is low and the experience paths are limited. The response is rather sluggish because of the formidable memory-capacity and processing-speed requirements. However, computers are becoming more powerful and less expensive all the time.

Expectation: Computer memory capacity increases by about 100 percent per year. The latest efforts toward development of *single electron memory*

(SEM) chips and biochips raise hopes of computers rivaling the human brain in terms of data density. Processing speed, too, keeps increasing, as clock speeds get faster and data buses get wider. Nevertheless, expectations in VR have historically run ahead of the technology.

Reactions: Some technophiles find VR so compelling that they use it as an escape from reality, rather than as an entertainment device. Proponents of VR argue that this does not reflect a problem with VR, any more than "computer addiction" represents a problem with computers. The trouble, say these researchers, is in the minds of people who are maladjusted to begin with. Other people are afraid of VR experiences; some VR illusions are as intense as hallucinations caused by drugs. Another problem results from the *uncanny valley* phenomenon, in which people become apprehensive around smart machines.

See also UNCANNY VALLEY THEORY and TELEPRESENCE.

VISION SYSTEM

One of the most advanced features of a mobile robot is the *vision system*, also called *machine vision*. There are several different designs; the optimum design depends on the application.

Components of a visible-light system

A visible-light vision system must have a device for receiving incoming images. This is usually a *vidicon* or *charge-coupled device* video camera. In bright light an *image orthicon* can be used.

The camera produces an analog video signal. For best machine vision, this must be processed into digital form. This is done by an *analog-to-digital converter (ADC)*. The digital signal is then clarified by *digital signal processing (DSP)*. The resulting data goes to the robot controller. The illustration is a block diagram of this scheme.

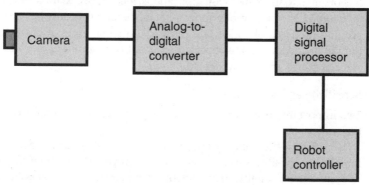

Vision system

The moving image, received from the camera and processed by the circuitry, contains an enormous amount of information. It is easy to present a robot controller with a detailed and meaningful moving image. It is more difficult to get a robot to "know" what is taking place in a particular scenario, based on the visual data it receives Processing an image, and getting all the meaning from it, is a challenge for vision-system engineers.

Vision and Artificial Intelligence

There are subtle things about an image that a machine will not notice unless it has an extremely advanced level of artificial intelligence (AI). How, for example, is a robot to "know" whether an object presents a threat? Is that four-legged creature a big dog, or is it a tiger? How is a robot to know the intentions of a moving object, if it has any? Is that biped creature a human being or another robot? Why is it carrying a stick? Is the stick a weapon? What does the biped want to do with the stick? It could be a woman with a closed-up umbrella, or a boy with a baseball bat. It could be an old man with a cane, or a hunter with a rifle. It is simple for a human being to tell the difference and to gauge the appropriate behaviors for dealing with any of these situations; programming a robot to have the same level of judgment is exceedingly complex. You know right away if a person is carrying a jack to help you fix a flat tire, or if the person is clutching a tire iron with which to smash your windshield. How is a robot to know things like this? It would be important for a police robot or a security robot to know what constitutes a threat, and what does not.

The variables in an image are much like those in a human voice. A vision system, to get the full meaning of an image, must be at least as sophisticated as a high-level *speech recognition* system. Technology has not yet reached the level of AI needed for human-like machine vision and image processing.

Fortunately, in many robot applications, it is not necessary for the robot to "comprehend" very much about what is happening. Industrial robots are programmed to look for certain easily identifiable things. A bottle that is too tall or too short, or a surface that is out of alignment, or a flaw in a piece of fabric, are easy to detect.

Sensitivity and resolution

Two important specifications in any vision system are the *sensitivity* and the *resolution*.

Sensitivity is the ability of a machine to see in dim light, or to detect weak impulses at invisible wavelengths such as *infrared* (IR) or *ultraviolet* (UV). In some environments, high sensitivity is necessary. In others, it is not needed and might not be wanted. A robot that works in bright sunlight does not need to be able to see well in a dark cave. A robot designed for working in mines, or in pipes, or in caverns, must be able to see in dim light, using a system that might be blinded by ordinary daylight.

Resolution is the extent to which a machine can differentiate between objects. The better the resolution, the keener will be the vision. Human eyes have excellent resolution, but machines can be designed with greater resolution. In general, the better the resolution, the more confined the field of vision must be. To understand why this is true, think of a telescope. The higher the magnification, the better will be the resolution (up to a certain point). However, increasing the magnification reduces the angle, or field, of vision. Zeroing in on one object or zone is done at the expense of other objects or zones.

Sensitivity and resolution depend somewhat on each other. Usually, better sensitivity means a sacrifice in resolution. Also, the better the resolution, the less well the vision system will function in dim light. Maybe you know this about photographic film. Fast film tends to have coarser grain, in general, than slow film.

Invisible and passive vision

Robots have a big advantage over people when it comes to vision. Machines can see at wavelengths to which we humans are blind.

Human eyes are sensitive to electromagnetic waves whose length ranges from approximately 390 to 750 nanometers (nm). The nanometer is a billionth (10^{-9}) of a meter. Light at the longest visible wavelength looks red. As the wavelength gets shorter, the color changes through orange, yellow, green, blue, and indigo. The shortest light waves look violet. Energy at wavelengths somewhat longer than 750 nm is IR; energy at wavelengths somewhat shorter than 390 nm is UV.

Machines need not, and often do not, see in the same range of wavelengths and the human eye sees. Insects can see UV that humans cannot, but are blind to red and orange light that humans can see. (Many people use orange "bug lights" when camping, or UV lamps with electrical devices that attract bugs and then zap them dead.) A robot can be designed to see IR or UV, or both, as well as (or instead of) visible light. Video cameras can be sensitive to a range of wavelengths much wider than the range humans can see.

Robots can be made to see in an environment that is dark and cold, and that radiates too little energy to be detected at any electromagnetic wavelength. In these cases the robot provides its own illumination. This can be a simple lamp, a laser, an IR device, or a UV device. Alternatively, a robot can emanate radio waves and detect the echoes; this is *radar*. Some robots can navigate via acoustic (ultrasound) echoes, like bats; this is *sonar*.

For further information

Comprehensive information can be found in a good college or university library. The best libraries are in the engineering departments at large universities. The Internet can also serve as a good source for information, but be sure to check the revision dates on all Web sites. Related definitions in this book include: BIN PICKING PROBLEM, BINOCULAR MACHINE VISION, BLACKBOARD SYSTEM, CHARGE-COUPLED DEVICE, COLOR SENSING, COMPOSITE VIDEO SIGNAL, COMPUTER MAP, DIRECTION RESOLUTION, DISTANCE RESOLUTION, EPIPOLAR NAVIGATION, EYE-IN-HAND SYSTEM, FLYING EYEBALL GUIDANCE SYSTEM, IMAGE ORTHICON, LADAR, LOCAL FEATURE FOCUS, LOG POLAR NAVIGATION, OBJECT RECOGNITION, OPTICAL CHARACTER RECOGNITION, PHOTOELECTRIC PROXIMITY SENSING, POSITION SENSING, PRESENCE SENSING, RADAR, RANGE SENSING AND PLOTTING, REMOTE CONTROL, RESOLUTION, SEEING-EYE ROBOT, SONAR, TELEOPERATION, TELEPRESENCE, TEXTURE SENSING, TRIANGULATION, VIDICON, and VIRTUAL REALITY.

VOICE RECOGNITION

See SPEECH RECOGNITION.

VOICE SYNTHESIS

See SPEECH SYNTHESIS.

VORONOI GRAPH

See GRAPHICAL PATH PLANNING.

W

WAYPOINT

See **METRIC PATH PLANNING.**

WELL-STRUCTURED LANGUAGE

A *well-structured language* is an advanced form of high-level computer programming language. These languages are used in *object-oriented programming,* such as is used in all personal computers, and also in robot-controller programming.

Assets

The main asset of a well-structured language is the fact that it can help a person write efficient, logical programs. Well-structured software can be changed easily. It often uses modular programming: programs within programs. Modules are rearranged and/or substituted for various applications. Well-structured programs lend themselves to easy debugging.

In most high-level languages, a computer program can be written in many different ways. Some are more efficient than others. The efficiency of a computer program can be measured in three ways, relative to the tasks the program is designed to carry out:

- The disk storage space required
- The memory required for the program to run
- The amount of computer time needed to run the program

These factors are closely correlated. An efficient program needs less storage, less memory, and less time to run than an inefficient one, when all other factors are held constant. When consumed memory and storage are minimized, the computer can access the data in minimal time. Thus it can solve the largest possible number of problems in a given length of time.

In *artificial intelligence* (AI), well-structured language is a requirement. In that field, the most demanding and complex in computer science, one must use the most powerful programming techniques available.

Two forms

Robot-controller program structuring can take either of two forms, which can be called *top-down programming* and *bottom-up programming*.

In the top-down approach, the computer user looks at the whole scenario, and zeroes in on various parts, depending on the nature of the problem to be solved. A good example of this is using a network to find information about building codes in Dade County, Florida. You might start with a topic such as State Laws. There would be a directory for that topic that would guide you to something more specific, and maybe even to the exact department you want. The programmer who wrote the software would have used a well-structured language to ensure that users would have an easy time finding data.

In the bottom-up approach, you start with little pieces and build up to the whole. A good analogy is a course in calculus. The first thing to do is learn the basics of algebra, analytical geometry, coordinate systems, and functions. Then, these are all used together to differentiate, integrate, and solve other complex problems in calculus. In a computerized calculus course, the software would be written in a well-structured language, so you (the student) would not waste time running into dead ends.

WHEEL-DRIVE LOCOMOTION

Wheel-drive locomotion is the simplest and cheapest way for a robot to move around. It works well in most indoor environments.

The most common number of wheels is three or four. A three-wheeled robot cannot wobble, even if the surface is a little irregular. A four-wheeled robot, however, is easier to steer.

The most familiar steering scheme is to turn some or all of the wheels. This is easy to do in a four-wheeled robot. The front wheels are on one axle, and the rear wheels are on another. Either axle can be turned to steer the robot. The illustration at the upper left shows front-axle steering.

Another method of robot steering is to run the wheels at different speeds. This is shown in the upper-right illustration for a three-wheeled robot turning left. The rear wheels are run by separate motors, while the front wheel is free-spinning (no motor). For the robot to turn left, the right rear wheel goes faster than the left rear wheel. To turn right, the left rear wheel must rotate faster.

A third method of steering for wheel-driven robots is to break the machine into two parts, each with two or more wheels. A joint between

One axle turns

Rear wheels run
at different speeds

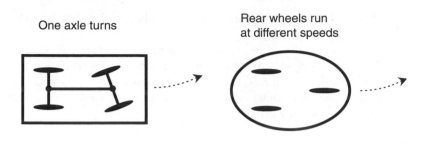

Two-section robot

Wheel-drive locomotion

the sections can be turned, causing the robot to change direction. This scheme is shown in the lower illustration.

Simple wheel drive has limitations. One problem is that the surface must be fairly smooth. Otherwise the robot might get stuck or tip over. This problem can be overcome to some extent by using *track-drive locomotion* or *tri-star wheel locomotion*. Another problem occurs when the robot must go from one floor to another in a building. If elevators or ramps are not available, a wheel-driven robot is confined to one floor. However, specially built tri-star systems can enable a wheel-driven robot to climb stairs.

Another alternative to wheel drive is to provide a robot with legs. This is more expensive and is more difficult to engineer than any wheel-driven scheme.

See also BIPED ROBOT, INSECT ROBOT, QUADRUPED ROBOT, ROBOT LEG, TRACK-DRIVE LOCO-MOTION, and TRI-STAR WHEEL LOCOMOTION.

WORK ENVELOPE

The *work envelope* is the range of motion over which a robot arm can move. In practice, it is the set of points in space that the *end effector* can reach.

The size and shape of the work envelope depends on the coordinate geometry of the robot arm, and also on the number of *degrees of freedom*. Some work envelopes are flat, confined almost entirely to one horizontal plane. Others are cylindrical; still others are spherical. Some work envelopes have complicated shapes.

The illustration shows a simple example of a work envelope for a robot arm using cylindrical coordinate geometry. The set of points that the end effector can reach lies within two concentric cylinders, labeled "inner limit" and "outer limit." The work envelope for this robot arm is shaped like a new roll of wrapping tape.

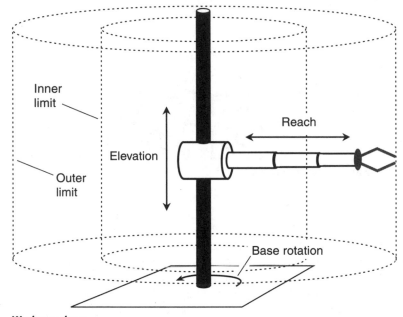

Work envelope

When choosing a robot arm for a certain industrial purpose, it is important that the work envelope be large enough to encompass all the points that the robot arm will be required to reach. But it is wasteful to use a robot arm with a work envelope much bigger than necessary. Compare CONFIGURATION SPACE and WORK ENVIRONMENT.

See also ARTICULATED GEOMETRY, CARTESIAN COORDINATE GEOMETRY, CYLINDRICAL COORDINATE GEOMETRY, DEGREES OF FREEDOM, DEGREES OF ROTATION, POLAR COORDINATE GEOMETRY, REVOLUTE GEOMETRY, ROBOT ARM, SPHERICAL COORDINATE GEOMETRY, and WORK ENVIRONMENT.

WORK ENVIRONMENT

The *work environment* of a robot, also called the *world space*, is the region in which a robot exists and can perform tasks. It differs from the *work envelope,* which represents the region of space that an *end effector* can reach when a robot is in a particular location.

In the case of a ground-based, mobile robot, the work environment can be defined in simplistic terms using a two-dimensional (2-D) coordinate system, specifying points on the surface such as latitude and longitude. With submarine or airborne mobile robots, the work environment is three-dimensional (3-D). Compare CONFIGURATION SPACE and WORK ENVELOPE.

WORLD MODEL

The term *world model* refers to the concept that a robot develops about its *work environment.* This concept is obtained from sensor outputs, previously obtained data (if any), and information the robot controller deduces concerning its optimum behavior. The world model should approximate physical and causative realities as closely as possible.

Every individual person has a concept of the environment—"the world around us"—but this differs slightly depending on various factors. In the same way, a robot's view of the world depends on factors such as

- The location of the robot
- The phenomena the robot can sense
- The sensitivity of the sensors
- The resolution of the sensors (if applicable)
- The computer map (if any) the robot controller has
- Information obtained from other robots
- Information obtained from humans
- The presence or absence of misleading input

Two identical robots in the same general location, and subjected to identical conditions, have identical world models unless one or both machines malfunctions, or one of the robots has a knowledge base that differs from that of the other. If the two robots have *artificial intelligence* (AI), and their "life experiences" differ, the robots can be expected to perceive the environment differently, even if they are in the same general location. Compare WORK ENVIRONMENT.

WORLD SPACE

See WORK ENVIRONMENT.

WRIST-FORCE SENSING

Several different forces exist at the point where a robot arm joins the *end effector.* This point is called the *wrist.* It has one or more joints that move in various ways. A *wrist-force sensor* can detect and measure these forces. It consists of specialized pressure sensors known as *strain gauges.* The

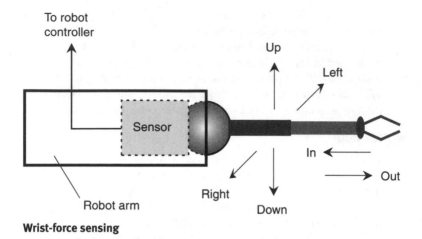

Wrist-force sensing

strain gauges convert the wrist forces into electric signals, which go to the robot controller. Thus the machine can determine what is happening at the wrist, and act accordingly.

Wrist force is complex. Several dimensions are required to represent all the possible motions that can take place. The illustration shows a hypothetical robot wrist, and the forces that can occur there. The orientations are right/left, in/out, and up/down. Rotation is possible along all three axes. These forces are called *pitch, roll,* and *yaw.* A wrist-force sensor must detect, and translate, each of the forces independently. A change in one vector must cause a change in sensor output for that force, and no others.

See also **BACK PRESSURE SENSOR, PITCH, PRESSURE SENSING, ROLL, TRANSDUCER, X AXIS, YAW, Y AXIS,** AND **Z AXIS.**

XYZ

X AXIS

The term *x axis* has various meanings in mathematics, computer science, and robotics.

In a *Cartesian plane* or *2-space* graph, the *x* axis is generally the horizontal axis (illustration, at left). The *independent variable* is represented by this axis. In Cartesian 3-space, the *x* axis is one of the two independent variables, the other usually being represented by *y* (illustration, at right).

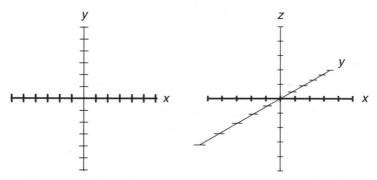

x axis

In *wrist-force sensing,* the *x axis* refers to linear forces from the right or the left. Compare Y AXIS and Z AXIS.

See also CARTESIAN COORDINATE GEOMETRY and WRIST-FORCE SENSING.

XR ROBOTS

The XR robots were manipulators conceived, designed, and built by a company called *Rhino Robots*. The main purpose of the devices was to demonstrate how robots work, and that there is no miracle involved in their functioning.

The XR robots were introduced during the 1980s, and sold for less than $3000 each. They did various tasks with high precision, and used a programming device similar to a *teach box*. For tasks involving numerous steps to be carried out in specific order, a personal computer could be used as the robot controller.

The XR robots proved useful as teaching aids in corporations and schools. Many people get uneasy around robots, especially the programmable type. The XR robots helped rid people of the fears they sometimes have about robots.

See also **EDUCATIONAL ROBOT, ROBOT ARM,** and **TEACH BOX.**

YAW

Yaw is one of three types of motion that a robotic *end effector* can make. Extend your arm out straight, and point at something with your index finger. Then move your wrist so that your index finger points back and forth (to the left and right) in a horizontal plane. This motion is yaw in your wrist. Compare **PITCH** and **ROLL.**

Y AXIS

The term *y axis* has various meanings in mathematics, computer science, and robotics.

In a *Cartesian plane* or *2-space* graph, the *y* axis is usually the vertical axis (illustration, at left). The *dependent variable* is represented by this axis. In a mathematical function f of an independent variable x, engineers specify $y = f(x)$. The function maps the x values into the y values.

In *Cartesian 3-space*, the *y* axis is one of the two independent variables, the other usually being represented by x (illustration, at right).

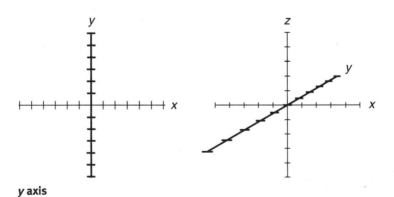

y axis

In *wrist-force sensing,* the y axis refers to inward/outward linear force vectors. Compare **x AXIS** and **z AXIS.**

See also **CARTESIAN COORDINATE GEOMETRY** and **WRIST-FORCE SENSING.**

Z AXIS

The term *z axis* has various meanings in mathematics, computer science, and robotics.

In *Cartesian 3-space,* the z axis represents the *dependent variable,* which is a function of x and y, the two *independent variables.* The z axis runs vertically, while the (x, y) plane is horizontal, as shown in the illustration. A function f maps values x and y into values z, such that $z = f(x, y)$.

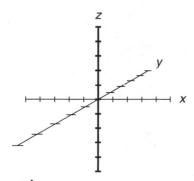

z axis

In *wrist-force sensing,* the z axis refers to up/down (vertical) linear force vectors. Compare **x AXIS** and **y AXIS.**

See also **CARTESIAN COORDINATE GEOMETRY** and **WRIST-FORCE SENSING.**

ZOOMING

In a robotic *vision system,* the term *zooming* refers to magnification of the image. If you want to look at a certain part of the screen in more detail, you can zoom in on it.

The illustrations show a hypothetical, infinitely complex shoreline or boundary. The lowest magnification is at the top left. Zooming in on a specific part of this graphic, more detail is revealed (top right). The zooming process is repeated, revealing still more detail (lower illustration). Because the boundary is irregular at all scales, the zooming can be done over and over indefinitely, and there is always new detail in the image.

Zooming, while increasing the magnification in theory, can increase the *resolution* only up to a certain point, depending on the quality of the optical system used. In general, the larger the lens diameter, the better is

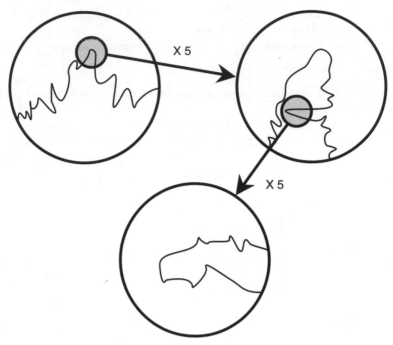

Zooming

the ultimate resolution. Zooming also limits vision by narrowing the field of view. In the examples shown, the view at top right has ⅕ (20 percent) of the angular diameter of the view at top left; the view at bottom has ⅕ (20 percent) of the angular diameter of the view at top right, and thus ¹⁄₂₅ (4 percent) of the angular diameter of the view at top left.

See also **VISION SYSTEM.**

Suggested Additional References

Arkin, Ronald C., *Behavior-Based Robotics*. Cambridge, Mass.: MIT Press, 1998.

Cook, David, *Robot Building for Beginners*. Berkeley, Calif.: APress, 2002.

Davies, Bill, *Practical Robotics*. Richmond Hill, Ontario: WERD Technology Inc., 2000.

Dudek, Gregory, and Jenkin, Michael, *Computational Principles of Mobile Robotics*. Cambridge, U.K.: Cambridge University Press, 2000.

Kortenkamp, David, Bonasso, R. Peter, Murphy, Rob, and Murphy, Robin R., *Artificial Intelligence and Mobile Robots*. Cambridge, Mass.: MIT Press, 1998.

Lunt, Karl, *Build Your Own Robot!* Natick, Mass.: A K Peters Ltd., 2000.

McComb, Gordon, *Robot Builder's Bonanza*. New York: McGraw-Hill, 2000.

Murphy, Robin R., *An Introduction to AI Robotics*. Cambridge, Mass.: MIT Press, 2000.

Sandler, Ben-Zion, *Robotics: Designing the Mechanisms for Automated Machinery*. Boston: Academic Press, 1999.

Winston, Patrick Henry, *Artificial Intelligence*, 3d ed. Addison-Wesley, 1992.

Wise, Edwin, *Applied Robotics*. Indianapolis, Ind.: Howard W. Sams, 1999.

Index

accent, 296–297
acoustic direction finder, 74
acoustic interferometer, 239
acoustic noise, 1
acoustic proximity sensor, 1–2
acoustic transducer, 292
active beacon, 25
active chord mechanism, 2, *2*, 230
active cooperation, 57
actual range, 114–115
actuator, 3, 39
adaptive robot, 267
adaptive suspension vehicle, 3
adhesion gripper, 3, 14
algorithm, 4, 192
aliasing, 32, 209
alkaline cell, 87, 89–90
all-translational system, 4
allophone, 296–298

alternative computer technology, 4–7
ampere hour, 88
amplitude modulator, 123
amusement robot, 7
analog computer technology, 5, 202
analog error accumulation, 96
analog image, 43
analog process, 5
analog-to-digital converter, 33, 39, 62, 85, 195, 240, 294, 337
analogical motion, 7–8, *8*, 71
analytical engine, 8–9, 10
AND, 35
android, 7, 9, 31–32, 111, 116, 120, 139, 195, 333
angular displacement transducer, 76
animism, 9–10
anthropomorphism, 10
antibody robot, 6, 153

Italicized numerals indicate illustrations.

Index

Apollo 11, 271
apparent range, 114–115
armature coil, 197–198
Armstrong, Neil, 272
array, 189
articulated geometry, 10–11, *11*, 99
artificial intelligence, 11–12, *12*, 51, 120, 139, 166–167, 190, 195, 202–203, 241, 260, 308, 323, 338
artificial stimulus, 12–13
articulated geometry, 164
ASCII, 297
Asimov, Isaac, 13
Asimov's three laws, 13, 117, 223
assembly robot, 13–14
atomic data, 6
attraction gripper, 14
attractive radial potential field, 231–232
automated guided vehicle, 12, 14–15, *15*, 93, 130
automated integrated manufacturing system, 13, 268
automation, 15
automaton, 15–16
autonomous robot, 76, 135, 136, 137, 268, 286, 307, 310
autonomous underwater vehicle, 304
axiom, 148
axis interchange, 16–17, *17*
axis inversion, 17–19, *18*
azimuth, 19–20, 53, 300
azimuth-range navigation, 19–20, *19*
azimuth resolution, 75

Babbage, Charles, 8, 10
back lighting, 21, 117, 284
back pressure, 21
back pressure sensor, 21–22, *22*, 25, 47, 81
back voltage, 21
backward chaining, 22
ballistic control, 22–23, 47

bandwidth, 23–24, *23*
bar coding, 13, 24, *24*, 30, 122, 211, 220
battery, 86
base 2, 206
base 10, 205
base rotation, 11, 59, 300
bat, 334
beacon, 24–25, 77, 78, 130
behavior, 25
Bell, Alexander Graham, 293
biased search, 25–26, *26*, 122
bin picking problem, 29, 128, 211
binary number system, 206
binary search, 26–27, *27*
binaural machine hearing, 27–28, *28*
binocular machine vision, 9, 28–29, *29*, 57, 159, 220
biochip, 6, 30–31
biological robot, 31, 188
biomechanism, 31
biomechatronics, 31–32
biped robot, 9, 32, 81
bird, 334
bit map, 32, 248
bit-mapped graphics, 32–33
bits per second, 23, 195
blackboard, 33
blackboard system, 33–34, *33*
bladder gripper, 34
bladder hand, 34
bleeder resistor, 237
body capacitance, 40
bogey, 292
Bongard problem, 34–35, *35*, 221
Boolean algebra, 35–36, 178
Boolean theorems, 36
bottom-up programming, 342
branch point, 36
branching, 36–38, *37*
Brooks, Rodney, 153
buffer, 63
bumper, 237

bundling, of optical fibers, 107
burn-in, 38
byte, 189

cable drive, 3, 39, 121
capacitive pressure sensor, 39–40, *40*
capacitive proximity sensor, 40–41, *41*, 245
capacitor-input filter, *235*
Capek, Karel, 260
card, 196
Cartesian coordinate geometry, 4, 16–19, 41–42, *42*, 121, 267
Cartesian coordinate system, 260
Cartesian plane, 347, 348
Cartesian 3-space, 348, 349
catastrophic failure, 126
CD-ROM, 5
central processing unit, 55, 192
centralized control, 42–43, 57
chain drive, 3, 43
charge-coupled device, 43–44, *44*, 337
checkers, 45
chess, 45
choke-input filter, *235*
chopping wheel, 216–217
cipher, 58
circuit board, 196
circuit breaker, 237
clean room, 45–46
clinometer, 46, *46*
clock frequency, 137
cloning, 31, 265
closed-loop configuration, 156
closed-loop control, 46–47
closed-loop system, 54, 104
coexistence, 47–48
cognitive fatigue, 48
cognizant failure, 48–49
color digital image, 70
color picture signal, 52
color sensing, 49–50, *50*

commutator, 125, 197–198
comparator, 97, 295
comparator IC, 157
complementary metal-oxide semiconductor, 50–51, 158
complex-motion programming, 51, *51*
compliance, 51–52
compliant robot, 51
composite video signal, 52–53, *52*, 226
computer-assisted instruction, 334–335
computerized axial tomography, 304
computer map, 53, 67, 108, 114, 123, 127, 130, 173, 192, 193, 261, 288, 307
configuration space, 53–54
conscious behavior, 25
constant-dropoff profile, 185
contact sensor, 54
context, 54, 295
continuous assistance, 283
continuous-path motion, 54–55, 283, 329
control trading, 56, 284
controller, 55–56, *55*
cooperation, 56–57
cooperative mobility, 57
correspondence, 57–58, *57*, 159
cosine wave, 68
Countess of Lovelace, 9
cryogenic technology, 204
cryptanalysis, 58
cybernetics, 58–59
cyberphobia, 325–326
cyborg, 59
cybot society, 59
cyclic coordinate geometry, 59
cylindrical cell, 90
cylindrical coordinate geometry, 59–60, *60*, 344

D'Arsonval meter, 92
data compression, 61
data conversion, 61–64, *62*, *64*

decentralized control, 78
decimal number system, 205–206
declining discharge curve, 88–89
dead reckoning, 214
deductive logic, 178
deductive reckoning, 214
definition, 264
degrees of freedom, 11, 13, 53, 64–65, 343
degrees of rotation, 65–66, 66
deliberation, 66–67, 141–142
deliberative paradigm, 138
deliberative planning, 67
demodulation, 195
depalletizing, 219
dependent variable, 348, 349
depth map, 67–68, 67
derivative, 68–69, 68, 69
desensitization, 327
destination, 61
Devol, George, 151
dialect, 297
dichotomizing search, 26
dielectric constant, 289
differential amplifier, 69–70, 69
differential transducer, 70
differentiator, 68
digital calculator, 8
digital error accumulation, 96–97
digital image, 43, 70–71
digital motion, 71
digital integrated circuit, 157–158
digital process, 4–5
digital signal processing, 63, 71–73, 72, 337
digital-to-analog converter, 63, 195, 297
direction finding, 73–75, 74, 99
direction resolution, 75, 173, 264, 292
directional transducer, 73
discharge curve, 88–89
displacement error, 75–76
displacement transducer, 76

distance measurement, 76–77, 114, 185, 316
distance resolution, 77–78, 264, 292
distinctive place, 78
distributed control, 78
domain of function, 79, 79, 118, 255
Doppler effect, 20
Doppler radar, 253
double integration, 214
downlink, 275, 326–327
Drexler, Eric, 199
drone, 95, 283
drop delivery, 80, 121, 140
dry cell, 87
dual-axis inversion, 19
duty cycle, 80–81
dynamic stability, 81
dynamic loudspeaker, 81
dynamic microphone, 81
dynamic RAM, 189
dynamic transducer, 81–82, 82

echo, 253, 316
edge, 261
edge detection, 83–84, 83, 130, 159, 288
educational robot, 84
elastomer, 40, 84–85, 85
electric eye, 85–86, 86, 238
electric generator, 91, 125
electric motor, 91, 197
electrically erasable programmable read-only memory, 190
electrochemical cell, 86
electrochemical power, 86–91, 87, 88, 89
electromagnetic field, 91
electromagnetic interaction, 149
electromagnetic interference, 91, 175
electromagnetic shielding, 91
electromechanical transducer, 91–92
electromechanics, 187
electron gun, 146–147, 330
electrostatic transducer, 92–93, 92

elevation, 11, 53, 59, 300
embedded path, 93
emitter-coupled logic, 157–158
empirical design, 93–94
end effector, 3, 14, 34, 39, 43, 80, 94, 97, 108, 140, 186, 226, 267, 270, 273, 293, 314, 329, 343, 345
endless loop, 182, 259
Engelberger, Joseph, 151
Enigma, 58
entitization, 94
epipolar navigation, 94–96, *95*, 108
erasable programmable read-only memory, 109, 190
error accumulation, 96–97
error correction, 97
error-sensing circuit, 97–98, *98*
error signal, 98–99, *99*
etching pattern, 240
event simulator, 285
exclusive OR gate, 179
exoskeleton, 99–100
expandability, 101
experience path, 331
expert system, 22, 100–101, *100*, 115, 280
exploratory VR, 331–333
exponential-dropoff profile, 185–186
extensibility, 101
extrapolation, 101
eye-in-hand system, 101–102, *102*, 108, 128

false negative, 103–104, 281
false positive, 103–104, 281
fault resilience, 104
feedback, 21, 47, 104–105, *105*, 143, 182, 215
fiber-optic cable, 105–107, *106*
fiber-optic data transmission, 5–6, 91
field of view, 107–108, *108*, 114
field coil, 197–198

field-effect transistor, 50
fifth-generation robot, 269–270
filter, 235
filter choke, 235
fine motion planning, 108, 128, 130
fire-protection robot, 108–109, 313
firmware, 109, 136, 190
first formant, 294
first-generation robot, 268–269
first-in/last-out, 259
fixed-sequence robot, 109
flash memory, 189
flat discharge curve, 88
flexible automation, 109–110
flight telerobotic servicer, 110–111, *110*
flip-flop, 189
flooded cell, 90–91
flowchart, 4, 111
fluxgate magnetometer, 111–112, *112*
flying eyeball, 112–113, *113*, 304
focus specialist, 34
foil run, 240
food-service robot, 113–114
force sensor, 34
foreshortening, 114–115, *115*
formant, 294
forward chaining, 115
four-phase stepper motor, 302–303
fourth-generation robot, 269–270
frame, 115–116, *116*
Frankenstein scenario, 116–117
frequency, 137
frequency-division multiplex, 198
frequency modulator, 123
front lighting, 21, 117
full-duplex module, 153
full-wave bridge rectifier, *234*
full-wave center-tap rectifier, *234*
full-wave rectifier, 234–235
fully centralized control, 43, 78
function, 55, 68, 79, 117–119, *118*
function generator, 119, 123

fuse, 237
futurist, 119–120
fuzzy logic, 178, 202

gantry robot, 121, 140
gas-station robot, 121–122, *122*
gateway, 122–123, 317
generator, 123–125, *124, 125*
geomagnetic field, 111
gigabits per second, 23
gigabyte, 189
gigahertz, 137
gimbal, 132
Global Positioning System, 126
goal node, 192, 203, 329
Gödel, Kurt, 148–149
graceful degradation, 104, 126–127,
   126, 201–202
graded-index optical fiber, 106
graphical path planning, 127–128, *127*,
   128
grasping planning, 128
gravity loading, 97, 129
grayscale, 49, 129
grayscale digital image, 70
gripper, 21, 80, 94, 101, 186
gross motion planning, 128, 129–130
groundskeeping robot, 130–131
group VR, 335
guidance system, 131
gyroscope, 4, 131–132, *131*

"Hacker" program, 133–134, *134*
"Hal," 10, 117
half-wave rectifier, 234, *234*
hallucination, 134–135
handshaking, 135–136, *136*
hard wiring, 136–137
hardware handshaking, 136
head-mounted display, 333–334
hertz, 137
heuristic knowledge, 137–138, 171

hexadecimal number system, 206
hierarchical paradigm, 138, 141, 257
high-level language, 138–139
hobby robot, 139, 223
hold, 139–140
holding, 139
home position, 140
household robot, 222, 287
human engineering, 140
hunting, 141, *141*
hybrid deliberative/reactive paradigm,
   141–142
hydraulic drive, 139, 142
hysteresis loop, 142–143, *143*

ideal battery, 88
ideal cell, 88
IF/THEN/ELSE, 145–146, *146*
ignorant coexistence, 47
image compression, 61
image orthicon, 146–147, *147*, 337
image resolution, 32, 48, 209, 226,
   291–292
immortal knowledge, 147–148, 171
incompleteness theorem, 148–149, *149*,
   242
independent variable, 229, 347, 348,
   349
individual VR, 335
inductive proximity sensor, 149–150,
   150, 245
industrial robot, 150–151
inertial guidance system, 131
inference engine, 100, 115, 151
infinite loop, 111, 182, 259
infinite regress, 151–152
inflection, 294, 296–298
informed coexistence, 47–48
infrared, 5
infrasound, 292
initial node, 192, 203, 329
input/output module, 152–153, *152*

insect robot, 16, 47, 77, 153, 203, 229, 258, 286, 327
instructional robot, 84
integral, 154–155, *154*
integrated circuit, 6, 30, 45, 50, 73, 109, 155–159, *157, 158, 159,* 192, 199, 236
intelligent coexistence, 47–48
intelligent mechatronic system, 267
intended function, 119
interactive simulator, 285
interactive solar-power system, 290
interactive VR, 331–333
interest operator, 158–159
interface, 159–160
interference pattern, 238
interferometer, 238
interferometry, 245
intermediate node, 192, 203
intermittent failures, 38
intermodulation, 327
interpolation, 160–161, *160, 161*
inverter, 179
inverting input, 156
ionization potential, 289
IR motion detector, 239
IR presence sensor, 238

jaggies, 32, 209
jaw, 163
join, 163–164
joint-force sensing, 164
joint-interpolated motion, 164–165, *165*
joint parameters, 164–165
joystick, 165–166, *166*
Jungian world theory, 166–167

K-line programming, 169–170, *170*
Karatsu, Hajime, 249
kilobits per second, 23, 195
kilobyte, 189
kilohertz, 137

kilowatt hour, 87
kinematic error, 76, 97, 169
kludge, 170–171
knowledge, 171
knowledge acquisition, 171
Kwo, Yik San, 304

ladar, 77, 173, 245, 316
ladle gripper, 173–174
landmark, 174, *174,* 317
landmark pair boundary, 174
lantern battery, 90
laser data transmission, 174–176, *175*
laser detection and ranging, 173, 245
laser radar, 173
latency, 48, 56, 283
lead-acid cell, 86–87, 90
Lecht, Charles, 166–167, 215
legged locomotion, 272
lidar, 173
light detection and ranging, 173
light-emitting diode, 216–217
line filter, 254–255
linear displacement transducer, 76
linear-dropoff profile, 185–186
linear integrated circuit, 156–157
linear interpolation, 160
linear programming, 176–177, *176*
lithium cell, 90
load/haul/dump, 177
local feature focus, 159, 177–178
Lockheed Aircraft, 322
log-polar transform, 180–181, *181*
logic, 178
logic equation, 36
logic families, 192
logic function, 118
logic gate, 30, 179–180, *180,* 199, 260
logic states, 4–5
look-ahead strategy, 45
loop, 181–182, 200
loop antenna, 75

lossless image compression, 61
lossy image compression, 61
Ludd, Ned, 182
luddite, 182

machine language, 4–5, 183, *183*
machine vision, 337
machining, 184
macroknowledge, 184
magnetic attraction gripper, 14
magnitude profile, 185–186, *185*
manipulator, 267
manually operated manipulator, 267
mapping, 117
master-slave manipulator, 311
mathematical induction, 178
maximum deliverable current, 88
mean time before failure, 186–187, *187*
mean time between failures, 186–187
mechatronics, 187–188
medical robot, 188–189, 195
megabits per second, 23, 195
megabyte, 189
megahertz, 137
memory, 189, 295
memory backup, 189
memory organization packet, 190
mercuric-oxide cell, 90
mercury cell, 90
message passing, 191, *191*
metal-oxide semiconductor, 191–192
metric path planning, 127, 192, 203
microcomputer, 192–193
microcomputer control, 193
microknowledge, 193
microphone, 92
microprocessor, 192
microwave data transmission, 193–194, *194*, 275
microwave presence sensor, 238
microwaves, 253
military robot, 195

mission planning, 141–142
mobile robot, 53, 127, 129, 130, 177
modem, 195–196, *196*
modular construction, 196
modular programming, 341
modulation, 195
module, 196
modulo 2, 206
modulo 10, 205
molecular computer, 6, 200
monocular vision, 28
motor, 125, 197–198, *197*
motor/generator, 125
multiagent team, 153
multiplex, 198
multiplexer IC, 156–157
Murphy's law, 104
mutual capacitance, 41

N-channel metal-oxide semiconductor, 158
NAND, 35
NAND gate, 179
nanochip, 199
nanorobot, 6, 199
nanorobotics, 199
nanotechnology, 6
natural language, 200
negation, 35
negative logic, 179
neighborhood, 78
nested loops, 200–201, *201*
nesting of loops, 181, 200–201
neural network, 7, 201–203
nickel-cadmium cell, 87, 90–91
nickel-metal-hydride cell, 91
node, 127, 203, 261, 280
noise, 203–204, *204*, 238
noise floor, 204
nonactive cooperation, 56–57
noninverting input, 156
nonservoed robot, 215

nonvolatile memory, 163, 189–190
NOR, 35
NOR gate, 179
NOT, 35
NOT gate, 179
nuclear service robot, 205
numeration, 205–208
numerically controlled robot, 267
Nyquist theorem, 62

object-oriented graphics, 33, 209–210, *210*
object-oriented programming, 341
object recognition, 30, 34, 94, 122, 201, 211, 222
objectization, 94
occupancy grid, 211–213, *212*
octal number system, 206
odometry, 213–215, *213*, *214*
offloading, 215
omnidirectional transducer, 73
one-dimensional range plotting, 256
one-to-one correspondence, 180
open-loop configuration, 156
open-loop system, 215–216
operational amplifier, 156
optical character recognition, 216, 297
optical encoder, 216–217, *217*
optical fiber, 105
optical presence sensor, 238
optical scanning, 216
optics, 5–6
OR, 35
OR gate, 179
orientation region, 174
orthogonal potential field, 232

P-channel metal-oxide semiconductor, 158
pallet, 219
palletizing, 219
parallax, 220, *220*

parallel data transmission, 63
parallel processing, 333
parallel-to-serial conversion, 63
partially centralized control, 42–43
partially distributed control, 78
passband, 294
passive transponder, 13, 30, 122, 220–221
passive VR, 331–33333
pattern recognition, 221
percept, 103, 107, 280
perpendicular potential field, 231–232
personal robot, 84, 109, 114, 170, 222–223, 268, 287
phase, 302
phase comparator, 27–28
phoneme, 223–224, 294, 296
photocell, 85
photodetector, 216–217
photoelectric proximity sensor, 224–225, *225*
photoreceptor, 146–147
photovoltaic cell, 289–290
piezoelectric transducer, 224–226, *226*
pitch, 53–54, 65, 132, 226, 346
pixel, 32, 43, 70–71, 129, 226–227
plan/act, 138
plan/sense/act, 138, 141, 257
pneumatic drive, 227
point-to-point motion, 227–228, *227*, 283, 302, 329
polar coordinate geometry, 228–229, *228*
police robot, 229–230
polymorphic robot, 230
position sensing, 97, 213, 230, 264
positive logic, 170
postulate, 148
potential field, 230–232, *231*
power supply, 232–237, *233*, *234*, *235*, *236*
power surge, 236

power transistor, 236
presence sensing, 237–239
pressure sensing, 240
primary cell, 87
primary colors, 49–50
problem reduction, 241–242, *241*
programmable manipulator, 267
programmable read-only memory, 190
proprioceptor, 242
prosodic features, 243, 295
prosthesis, 31, 59, 99, 243–244
prototype, 93
proximity sensing, 40–41, 127, 130, 185,
    244–245, *245*
pushdown stack, 259

quadruped robot, 247–248, *248*
quadtree, 248–249, *248*
quality assurance and control, 249–251,
    250, 262

radar, 20, 53, 77, 211, 245, 253–254, *254*,
    316, 339
radiant heat, 239
radiant heat detector, 239
radio detection and ranging, 253
radio direction finding, 75, 97–98
radio-frequency interference, 254–255
radix 2, 206
radix 10, 205
Rand Corporation, 45
random-access memory, 55, 189
range, 19–20, 53, 67, 255
range image, 67
range of function, 118, 255–256, *256*
range plotting, 255–257, *257*
range sensing and plotting, 67, 173, 264
ranging, 76, 316
raster, 330
raster graphics, 32
reach, 11, 59, 300
reactive behavior, 25

reactive paradigm, 141, 257–258
reactive planning, 67
read-only memory, 109, 190
read/write memory, 33, 189
real time, 258
real-time failures, 38
rectangular coordinate geometry, 41
rectifier, 234–235
recursion, 258–260, *259*
reductionism, 260
reflexive behavior, 25
refractive index, 105
refresh rate, 48
regular grid, 260–261, *261*
reinitialization, 260–261
relation, 280
relational graph, 261, *262*
reliability, 261–263, *263*
remote control, 263–264, 271
remotely operated vehicle, 304
representation, 192
repulsive radial potential field, 231–232
resolution, 70–71, 264–265, *265*,
    338–339, 349–350
reverse engineering, 265–266
revolute geometry, 266, *266*, 267
revolutions per minute, 197
revolutions per second, 197
Rhino Robots, 347
robot arm, 3, 10–11, 65–66, 94, 139,
    186, 267, 299
robot car, 83
robot classification, 267–268
robot generations, 268–270
robot gripper, 270
robot leg, 272, 319
robot mouse, 7
robotic paranurse, 188
robotic ship, 270–271
robotic space travel, 271–272
roll, 53–54, 65, 132, 273, 346
Rossum's Universal Robots, 260

"Rube Goldberg" contest, 171
rule-based system, 100

sabotage-proof system, 104
sampling interval, 62
sampling rate, 62
sampling resolution, 62, 264
sampling theorem, 62
satellite data transmission, 275–276, *275*
scaling, 276–277, *277*
second formant, 294
second-generation robot, 268–269
secondary cell, 87
secondary electron, 146
security robot, 277–278, 287
seeing-eye robot, 278
selsyn, 92, 279, *279*, 305
semantic network, 279–280
sense/act, 257–258
sensitivity, 338–339
sensor competition, 280–281
sensor fusion, 281
sentry robot, 281–282, 287
sequential manipulator, 267
serial data, 135
serial data transmission, 63
serial-to-parallel conversion, 63
servo, 74
servo robot, 283
servo system, 283
servomechanism, 47, 92, 97, 102, 215, 282–283, 293
shape-shifting robot, 230
shared control, 56, 283–284
shelf life, 88
side lighting, 21, 117, 284
signal comparison, 73–74
signal generator, 123–125
signal-to-noise ratio, 63
simulation manager, 331
silver-oxide cell, 90

simple-motion programming, 284–285, *285*
simulation, 284–286
sine wave, 68
single-axis inversion, 18
single-electron memory, 6, 336–337
smart home, 49, 286–288
smart robot, 267, 268
smoke detection, 286, 288–289, *289*
society, 153
software handshaking, 136
solar cell, 289–290
solar power, 289–290
sonar, 1, 53, 77, 114, 211, 245, 290–292, *291*, 316, 339
sound detection and ranging, 290
sound system, 334
sound transducer, 27–28, 292, 334
source, 61
spacecraft cell, 91
spatial resolution, 264, 292–293
speaker, 92
spectrum space, 23
speech recognition, 9, 34, 54, 81, 93, 139, 201, 222, 243, 293–296, *293*, 298–299, 305, 334
speech synthesis, 9, 81, 93, 139, 222, 243, 296–299, *298*, 305, 334
spherical coordinate geometry, 299–300, *299*
spherical coordinates, 53
stadimetry, 300–301, *300*
stand-alone solar power system, 290
standing waves, 1
static RAM, 189
static stability, 301
step angle, 301
step-down transformer, 233, *233*
step-index optical fiber, 106
step-up transformer, 233, *233*
stepper motor, 3, 71, 92, 139, 301–303, *303*

storage capacity, 87–88
storage media, 55
strain gauge, 345–346
subject/verb/object, 305
submarine robot, 112, 302–304
surge suppressor, 236, 254
surgical assistance robot, 304
Sussman, Gerry, 133
swarm, 153
synchro, 304–305
syntax, 295, 305

tactile sensation, 48
tactile sensing, 84, 108, 128, 211, 307
tangential potential field, 231–232
task environment, 307–308
task-level programming, 26, 308, *308*
teach box, 56, 283, 308–309, 348
technocentrism, 309–310
technophobe, 181
telechir, 310–313, 333
telemetry, 310
teleoperation, 56, 205, 271, 275, 283,
    304, 310–311
telepresence, 48, 109, 111, 205, 263,
    271–272, 275, 303, 311–313, *311*,
    333, 335–336
temperature sensing, 313–314, 315
terabyte, 189
tethered robot, 314
texture sensing, 314–315, *314*
theorem-proving machine, 241
thermistor, 315
thermocouple, 315–316
Theta Tau, 171
third formant, 294
third-generation robot, 268–269
three-dimensional range plotting,
    256–257
threshold-detection profile, 185
tic-tac-toe, 45
time-division multiplex, 198

time-of-flight distance measurement, 316
time-of-flight ranging, 316
time sharing, 258
time shifting, 316–317
timer IC, 156
top-down programming, 342
topological navigation, 317
topological path planning, 122, 317
torque, 302
total internal reflection, 106
track drive, 9, 139
track-drive locomotion, 317–319, 318,
    343
transducer, 319
transformer, 233–234
transient, 236
transient suppressor, 236, 254
transistor battery, 90
transistor-transistor logic, 157–158
tri-star wheel locomotion, 318, 319,
    321–322, *321*, 343
triangulation, 320–321, *320*
tricorner reflector, 320
trinary logic, 178
triple-axis inversion, 19
truth table, 322–323
Turing, Alan, 58, 323
Turing test, 323
two-dimensional range plotting, 256
two-phase stepper motor, 302–303
two-pincher gripper, 323–324, *324*

ultrasonic motion detector, 239
ultrasound, 290–291, 292, 304
ultraviolet, 5
uncanny valley phenomenon, 337
uncanny valley theory, 325–326, *326*
uniform potential field, 231
uniformly distributed control, 78
uninterruptible power supply, 236
uplink, 275, 326–327
user friendliness, 140

vacuum cup gripper, 329
vector, 230–232
vector array, 230
via point, 228, 329
videocassette recorder, 330
vidicon, 330, *330*, 337
virtual reality, 271, 331–337, *332, 333,
    335*
virtual universe, 331
virtual virtual reality, 331
visible light, 5
vision system, 32, 49, 53, 83, 95, 130,
    139, 177–178, 211, 216, 222, 264,
    290, 337–340, *337,* 349
voice recognition, 293
voice synthesis, 296
volatile memory, 189
voltage regulator, 236, *236*
voltage-regulator IC, 156
Voronoi graph, 127–128
Voyager, 271, 310

Wasubot, 9
watt hour, 87
waypoint, 192, 203

well-structured language, 341
wheel drive, 9, 39, 139, 342–343
wheel-drive locomotion, 342–343, *343*
whiskers, 237–238
work envelope, 16, 65, 308, 343–344,
    344, 344–345
work environment, 174, 248, 292,
    344–345
world model, 345
world space, 53, 248, 307, 344
wrist, 345–346
wrist-force sensing, 345–346, *346,* 347,
    349

x axis, 347, *347*
XOR gate, 179
XR robot, 347, 348

y axis, 348–349, *348*
yaw, 53–54, 65, 132, 346, 348

z axis, 349, *349*
Zener diode, 236
zinc-carbon cell, 87, 89
zooming, 349–350, *350*

About the Author

Stan Gibilisco has authored or coauthored dozens of nonfiction books about electronics and science. He first attracted attention with *Understanding Einstein's Theories of Relativity* (TAB Books, 1983). His *Encyclopedia of Electronics* (TAB Professional and Reference Books, 1985) and *Encyclopedia of Personal Computing* (McGraw-Hill, 1996) were annotated by the American Library Association as among the best reference volumes published in those years. Stan serves as Advisory Editor for the popular *Teach Yourself Science and Mathematics* book series published by McGraw-Hill. His work has gained reading audiences in several languages throughout the world.